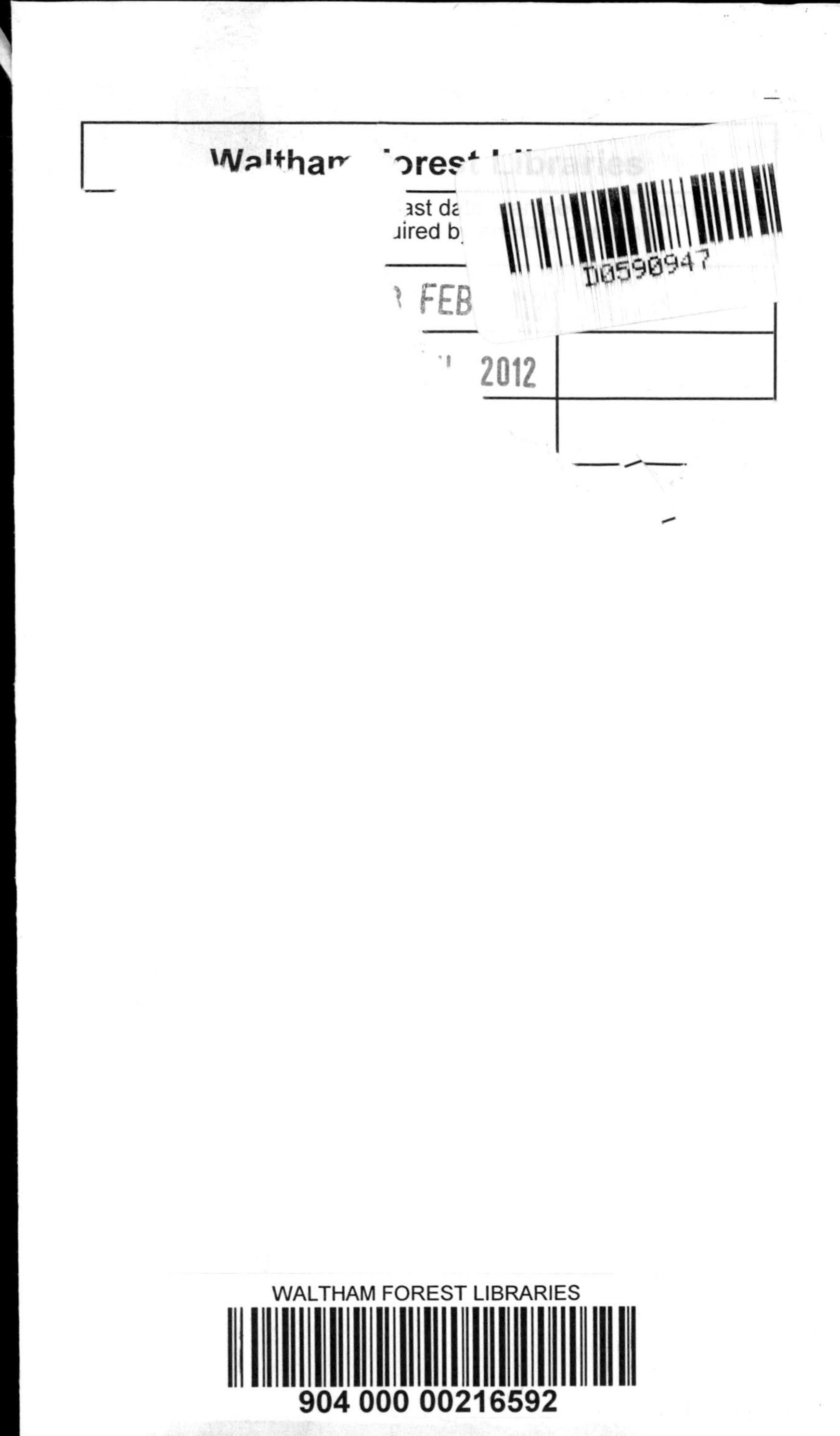

THE CALCULUS DIARIES

THE CALCULUS DIARIES

How Math Can Help You Lose Weight, Win in Vegas, and Survive a Zombie Apocalypse

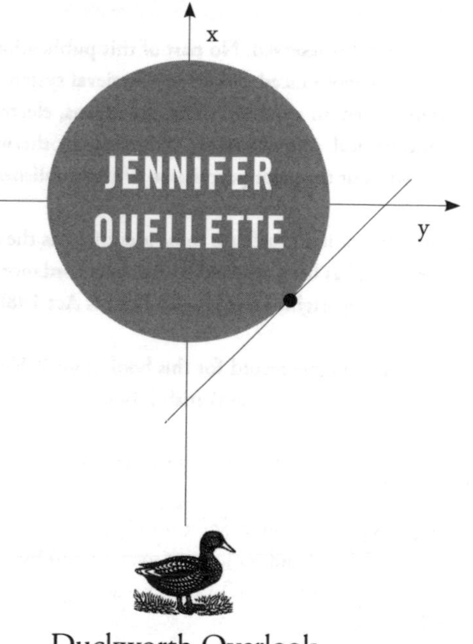

Duckworth Overlook

First published in the UK in 2011 by
Duckworth Overlook
90-93 Cowcross Street, London EC1M 6BF
Tel: 020 7490 7300
Fax: 020 7490 0080
inquiries@duckworth-publishers.co.uk
www.ducknet.co.uk

First published in the USA in 2010 by
the Penguin Group, New York

Figures in Appendix 1 by Sean Carroll

ISBN 978 0 7156 4143 9

Book design by Judith Stagnitto Abbate / Abbate Design
Printed and bound in Great Britain by
CPI Group (UK) Ltd, Croydon, Surrey

For Sean, the sine to my cosine.

Neglect of mathematics works injury to all knowledge, since one who is ignorant of it cannot know the other sciences, or the things of this world. And what is worst, those who are thus ignorant are unable to perceive their own ignorance, and so do not seek a remedy.

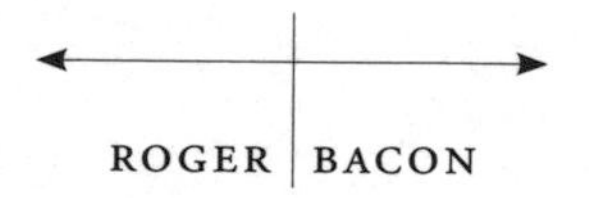

CONTENTS

ACKNOWLEDGMENTS

This book grew out of my impulsive Internet purchase of a DVD lecture series, *Calculus Made Clear*, offered by the Teaching Company. The instructor, a math professor at the University of Texas–Austin, named Michael Starbird, combined the necessary rigor—in the form of simple diagrams and derivations—with an appealing (to me) conceptual approach, punctuated with colorful historical anecdotes. Nothing makes an English major happier than a compelling narrative. Tell us a good story and we'll follow you anywhere, even into the minefield of scary calculus equations. Starbird's lectures inspired me first to write a series of blog posts on Cocktail Party Physics about my adventures exploring calculus, and then to expand those into a full-length book.

But Starbird had the advantage of working with fertile ground. I owe a great debt to Alan Chodos, a physicist who taught for years at Yale University before becoming associate executive officer of the American Physical Society. That gift for teaching never really left him. Alan not only encouraged me to write my first book, but carefully explained many basic physics concepts to me and insisted that I let him walk me through the relevant equations. We now live on opposite sides of the country, but Alan has had a lasting impact on my life, in the tradition of great teachers and mentors everywhere.

I also benefited greatly from close readings of a handful of other works, most notably Charles Seife's *Zero: The Biography of a Dangerous Idea*; Jason Bardi's *The Calculus Wars*; David Berlinski's *A Tour of the Calculus*; and Leonard Mlodinow's *The Drunkard's Walk*. All were instrumental in shaping my thinking about the concepts of calculus. When it came to putting those concepts into practice, I found W. Michael Kelley's *The Complete Idiot's Guide to Calculus* to be a helpful resource.

Thanks are due to the many people (mathephiles and mathephobes alike) who generously shared their stories and insights over the two years I spent researching and writing this book, including (but not limited to) Bisi Agboola, Dave Bacon, Jason Bardi, Allyson Beatrice, Adam Boesel, Ben Carey, Deborah Castleman, Rob Chiappetta, Calla Cofield, K. C. Cole, Julianne Dalcanton, Geoffrey Edelstein, Adam Frank, Milton Garces, David Grae, David Gross, Lauren Gunderson, Kevin Hand, David Harris, Joanne Hewett, Karen Heyman, Daniel Holz, Alice Hung, Valerie Jamieson, George Johnson, Rich Kim, Lee Kottner, Tom Levenson, M. G. Lord, Gabrielle Lyon, Malcolm MacIver, Alex Morgan, Chad Orzel, Dennis Overbye, Phil Plait, Joe Polchinski, Lisa Randall, Abbas Raza, James Riordon, David Saltzberg, Robert Smith?, Tara Smith, Shari Steelsmith-Duffin, Ben Stein, Brian Switek, Carol Tavris, Kip Thorne, Mark Trodden, Jatila van der Veen, Robin Varghese, Rosie Walton, Gordon Watts, Margaret Wertheim, Risa Wechsler, Glen Whitman, Carolee Winstein, Mark Wise, and Tony Zee. Extra special thanks to Janet Blumberg, Diandra Leslie-Pelecky, and Eric Roston, who slogged through parts of the draft manuscript and offered helpful critiques.

Jason Torchinsky did a fantastic job devising nifty illustrations of the abstract concepts throughout the book. Thanks are also due to Thomas Roberge, my editor at Penguin, and to my agent, Mildred Marmur, who offered her usual unstinting support and sage advice. And as always, I am deeply grateful to friends and family; miraculously, everyone still speaks to me after yet another lengthy disappearing act to write a book.

I took great pains to ensure I understood the underlying concepts, not just the mechanical processes of calculus. Invariably, this means channeling one's inner three-year-old and constantly asking "Why?" That can get pretty annoying. So I owe the greatest debt to my husband, Sean Carroll (aka the World's Most Patient Man), who put up with my inner toddler for two years. I'm not sure I would have written this book without him. He helped me "find the calculus" in each chapter and allowed me to caricature him in the text for comic effect. He also proofread the entire manuscript and deftly avoided the odd bit of metaphorical heaved crockery when I hit an obstacle ("integrate *that*!"), calmly guiding me toward the solution. I promise my next book will be about something simpler, like butterflies and rainbows. Or bunnies. Surely there's no math in bunnies.

THE CALCULUS DIARIES

PROLOGUE

I Could Be Mathier

> **Xander:** Giles lived for school. He's actually still bitter that there are only twelve grades.
>
> **Buffy:** He probably sat in math class thinking, *There should be more math. This could be mathier.*
>
> —"THE DARK AGE," *Buffy the Vampire Slayer*

Archimedes of Syracuse was the quintessential math nerd. Granted, he invented many practical devices, including devastatingly effective engines of war that helped Syracuse beat back an attack by the Roman general Marcellus in the siege of 212 B.C.—at least temporarily. But his one true love was pure mathematics, especially geometry. The Roman historian Plutarch tells how Archimedes' servants had to forcibly bathe their preoccupied master, who would sketch geometrical figures in chimney embers, and in the oils that anointed his naked body after bath time.

That single-minded obsession proved to be his downfall. Eventually Marcellus overcame Archimedes' ingenious defenses,

and Roman soldiers swarmed through the city of Syracuse. Historical accounts report that Archimedes was so engrossed in studying a geometric figure he'd drawn in the dust that he barely noticed the chaos around him. A Roman soldier "in quest of loot" marched up to the scholar and demanded that Archimedes accompany him to Marcellus's tent. Archimedes demurred, saying he wished to finish solving his geometrical problem first: "I beg you, don't disturb this." Incensed, the soldier summarily killed him, so that "with his blood he confused the lines of his art."*

* This account is given by Valerius Maximus, in *Memorable Doings and Sayings*. Historians differ as to how the soldier slew Archimedes, but a medieval woodcut depicts his head being cleft in two. Several accounts report that Marcellus was much distressed by the mathematician's death, since he had great respect for the man's ingenuity—even though that ingenuity had delayed his conquering of Syracuse.

This account of the death of Archimedes provided inspiration centuries later, when a young French girl named Sophie Germain read the story in the late eighteenth century. She concluded that if someone could be so consumed by a geometric problem, then geometry must be the most fascinating subject in the world. So Germain set out to learn it, defying her family's strictures by studying math in secret under the bedclothes at night. Later she masqueraded as a male student at the École Polytechnique in Paris (girls were not admitted), and by the time she died of breast cancer in 1831, she was a highly accomplished mathematician.*

The soldier who killed Archimedes wasn't quite so inspired. Perhaps Archimedes reminded him uncomfortably of his high school math teacher, who may have ridiculed the soldier's failure to grasp the fundamentals of geometric proofs in front of snickering classmates. All that pent-up resentment and frustration boiled over into an impulsive act of rage, making the Greek scholar an early casualty in the longstanding war between jocks and nerds.

Pure conjecture, naturally, but many of us can relate—even more so when we learn that Archimedes came dangerously close to inventing calculus. Two thousand years later, traumatic memories of high school calculus evoke powerfully negative reactions among people of all ages, genders, and backgrounds. Most people would rather be strung up by their thumbs and systematically tortured with sharp, pointy objects than be forced ever again to find the antiderivative of a polynomial.

* Sophie Germain is best known for inventing the "Germain primes." If you double a Germain prime number and add 1, you get another prime number. For example, double the prime number 2 is 4, plus 1 is 5—which is also a prime number.

Math in general, and calculus in particular, is something to be avoided like the plague once we leave high school. An episode of the TV series *House* opens with a group of students taking the AP calculus exam. A boy collapses and is rushed to the hospital. When House is told of the circumstances of the boy's collapse, he quips, "That's the way calculus presents."

So calculus has a formidable reputation. I have always been among those nonmathematical sorts who viewed it with trepidation and preferred to keep a safe distance. In fact, I avoided taking calculus altogether by cleverly skipping out on my senior year of high school for early admission to college. Since I am a science writer who specializes in physics topics, it surprises many people to learn that I have a lingering phobia about math. Chalk it up to my English-major roots, but the sight of even a simple algebraic equation still elicits an involuntary shudder, unless I consciously counteract it.

I am not alone in my ambivalence. My friend Allyson, in particular, seems to be a kindred spirit to that long-ago Roman soldier. "My initial reaction to the word *calculus* is not unlike a caveman throwing rocks at the moon in ignorance and fear resulting in blind rage," she confessed when I asked about her aversion to all things math. "There is no such thing as ghosts creeping up behind me on the stairs, but there is such a thing as a polynomial monster, and it has hooked teeth and causes chronic yeast infections, I'm sure."

Our stubborn resistance to calculus is not entirely rational. Frankly, most of us don't even know what calculus entails; its reputation for being difficult and unpleasant precedes it. Calculus is quite simple and straightforward in concept; the devil is in the details. Essentially it's a way of measuring change, whether it be change in position, temperature, or what have you. Its power comes from its universality: The same basic concepts can be applied to systems as diverse as a car driving down a road, the stock market, the Black Death, or surfing. That's why calculus textbooks are so thick.

Calculus boils down to two fundamental ideas: (1) the *derivative* (differential calculus), which is a way of measuring instantaneous change, such as finding the speed of a car when you only know its position; and (2) the *integral* (integral calculus), which describes the accumulation of an infinite number of tiny pieces that add up to a whole and can be used, for instance, to determine the distance a car has traveled when only its speed is known. Everything else is just a variation on these two themes. The derivative and the integral are like the two ends of a hammerhead: One is for pulling out the nails, and the other is for pounding them in. The first is a process of subtraction and division; the second, a process of multiplication and addition. Each "undoes" the work of the other. And not every math problem requires a hammer; sometimes a screwdriver works best. So calculus is just one tool in a broad arsenal of mathematical instruments, applicable to specific kinds of problems.

I explained this to Allyson, who responded with an incredulous, "That's it? Why can't math teachers just *say* that?" In fairness, they probably do; they just say it in a foreign language. Galileo famously observed, "Nature's great book is written in mathematical symbols." Unfortunately, to the untrained eye and ear, that language resembles ancient Sanskrit, and math teachers may as well be speaking gibberish. Most of us never get past the strange symbols and jargon, and thus meander through life without any quantitative tools beyond basic arithmetic. We can balance a checkbook, but have no grasp of statistics, compound interest, or probability, for example—and this puts us at the mercy of those who *do* understand them, and thus can manipulate us at will. Knowledge is power, and we forfeit that power when we choose to remain willfully ignorant.

The inability to grasp basic algebra and calculus also can be a stumbling block to many students who otherwise would wish to become scientists. Take my friend Lee, whose struggles with algebra in high school—despite top grades in all her other classes—reduced her to tears, and kept her from becoming a marine biologist. She still loves science, but has a visceral hatred of mathematics to this day. "It wrecked my self-confidence in a way nothing else ever did, and still knots my stomach," she told me. "I'm not totally innumerate, but anything that looks like an equation makes me break out into a cold sweat and run screaming in the other direction."

Ironically, given my distaste for the subject, I succeeded at math, at least by the usual evaluation criteria: grades. Yet while I might have earned top marks in geometry and algebra, I was merely following memorized rules, plugging in numbers and dutifully crunching out answers by rote, with no real grasp of the significance of what I was doing or its usefulness in solving real-world problems. Worse, I knew the depth of my own ignorance, and I lived in fear that my lack of comprehension would be discovered and I would be exposed as an academic fraud—psychologists call this "impostor syndrome."

I might have gone through the rest of my life cringing compulsively at the mere sight of an equation. But I became a science writer and fell in love with physics—not the math part, mind you. I loved the rich history, the people, the funky experiments, and the big ideas. One fateful day, I asked a physicist named Alan *why* it was true that all objects fall at the same rate, regardless of mass, when casual observation would seem to indicate the opposite. It seemed counterintuitive to me.

This is the basis of a famous experiment proposed by Galileo. If you drop a coin and a feather under normal (atmospheric) conditions, the coin will hit the ground first. But Galileo reasoned that another force—air resistance—slowed down the feather's descent because it had more surface area than the coin. In a vacuum, there would be no air resistance, so all objects would accelerate equally. He didn't have the techniques for creating a vacuum back then to test his hypothesis, but Isaac Newton derived Galileo's assertion mathematically in the seventeenth century.

Today, vacuum technology is commonplace, and the coin-and-feather experiment is a staple of physics demonstrations. I had witnessed one such demonstration, so I knew from experience it was true that objects fall at the same rate regardless of mass—or did I? I hadn't built and performed the experiment myself. How could I know it wasn't some kind of trick, or a mistake in the experimental setup?

Alan pondered a moment, stroking his beard, and then pointed out that I need not take the matter on faith. It would become obvious to me why this was so if I allowed him to walk me through the equation.

I resisted. Alan persisted: "It's not *real* math; it's just algebra." He wore me down eventually, and he was right. On his office whiteboard, he patiently demonstrated how the little m—for the mass of the object, as distinct from a big M for the mass of the Earth—on each side of the equation effectively cancels out, making an object's mass irrelevant to the rate of acceleration. And I had my first epiphany that math might actually be relevant to my life: Among other advantages, mathematics can help us better grasp the more counterintuitive notions in physics. It is certainly possible to do so without the benefit of algebra or calculus, but when I saw that equation worked out, some final piece of insight clicked into place that I hadn't even realized was missing. Math finally had a meaningful context.

So began a gradual lowering of my knee-jerk defenses against numbers and abstract symbols, and the start of a grudging appreciation for the role of mathematics in the "real world." It was a tantalizing glimpse into a whole new way of looking at reality, and for the first time in my mathephobic life, I wanted to learn more. Armed with a few books,* a DVD lecture series from the Teaching Company, and the support of my physicist spouse, I set out to discover what I'd been missing all those years.

Once I started delving into calculus, I realized that this seemingly arcane subject is applicable to everything from gas mileage, diet and exercise, economics, and architecture to population growth and decline, the physics behind the rides at Disneyland, the probabilities associated with shooting craps in a Vegas casino—even the I Ching. In fact, one could argue that we all do some form of calculus all the time, without realizing it. A baseball outfielder has to estimate where the ball is likely to land after the batter hits it. Whether he knows it or not, his brain is calculating the trajectory of that ball, then sending a signal telling the outfielder where to place himself in order to

* To give you an idea of the depth of my ignorance at the outset, *The Complete Idiot's Guide to Calculus* proved to be a bit over my head. Perhaps it should be retitled *The Half-Wit's Guide to Calculus*.

make the catch. Lurking somewhere in that process is a calculus problem. Or two.

Even lowly worms do calculus, according to a University of Oregon biologist named Shawn Lockery. He's studied roundworms to figure out how they use their sense of taste and smell to navigate as they forage for food. He compares the approach to the game of hot-and-cold one might play with a child, in which one says, "You're getting warmer (or colder)" to help said child home in on the target. Roundworms do this, too, changing direction in response to feedback, but they get their feedback by calculating how much the strength of different tastes—in this case, salt concentrations—is changing. In calculus terminology, the worms take a derivative to figure out how much a given quantity is changing at a certain point in space and time, and adjust their behavior accordingly.

If worms can do calculus, human beings simply have no excuse for avoiding it. I think scientists have a valid point when they bemoan the fact that it's socially acceptable in our culture to be utterly ignorant of math, whereas it is a shameful thing to be illiterate. We could all be just a little bit mathier. We don't all need to become mathematical prodigies, but we ought to have some basic understanding of how math in general, and calculus in particular, fits into our cultural framework, and be able to look at a rudimentary equation without breaking into a cold sweat. It is an integral part of our intellectual history, after all.

William Benjamin Smith, a math professor in the late nineteenth century, observed in the preface to his book *Infinitesimal Analysis*, "Calculus is the most powerful weapon of thought yet devised by the wit of man." Far from being some static, dead set of rules to be memorized and blindly followed, calculus is almost an organic entity. Watch any physicists work a problem, and you'll see its extraordinary flexibility: They play fast and loose with the numbers, simplifying and rounding up as needed to complete the task at hand. They adapt calculus to their needs—not the other way around. The act of devising a calculus problem from your observations of the world around you—and then solving it—is as much a creative endeavor as writing a novel or composing a symphony. Those things are not easy, nor should they be. As with any art form, the best way to learn and improve is by diligently practicing that art.

In college, I proudly flaunted my mathematical ignorance by sporting a T-shirt reading, "English major—*you* do the math." I never realized, until much later, that this defensive, belligerent attitude stood in the way of acquiring genuine understanding. I have a new T-shirt now to symbolize my change in mindset: "You mess with calculus, you mess with me." Archimedes certainly felt that way about his geometric diagram. Mathematics, he knew, was universal, eternal, and to his mind, far more precious than life.

1

To Infinity and Beyond

> You take a function of x and you call it y,
> Take any x-nought that you care to try,
> Make a little change and call it delta-x,
> The corresponding change in y is what you find nex',
> And then you take the quotient and now carefully
> Send delta-x to zero, and I think you'll see
> That what the limit gives us, if our work all checks,
> Is what we call *dy*/*dx*, it's just *dy*/*dx*.
>
> —TOM LEHRER,
> "The Derivative Song"

You never know what you'll find moldering in a musty old attic: forgotten photo albums, moth-eaten vintage clothing, discarded toys—or maybe a rare mathematical manuscript disguised as a humble prayer book. That's what one French family discovered in the closet one day in the late 1990s: a battered and smudged prayer book with the faint outlines of Greek lettering in the margins, along with an occasional diagram. Sensing a potentially significant find, the family brought the book to Christie's auction house of London for appraisal. It proved a financially astute move: in 1998 the prayer book sold for $2 million.

What was so special about a tattered ancient prayer book? It took numerous scientists, digital photography under different wavelengths of light, and a spot of x-ray fluorescence imaging to fully decipher the mystery.* Almost a decade of intensive scientific analysis revealed that lying just under the surface text of those prayers are the scribblings of Archimedes. Not just random musings, either: It was two lost texts by the great mathematician, including one, entitled *The Method*, that constitutes the earliest known written work on what would later develop into integral calculus.

Archimedes wrote *The Method* on a scroll of papyrus over two thousand years ago. Eventually someone copied the text onto animal skin parchment and then set the copy aside to molder in a library in Constantinople until 1229 A.D. It was standard practice in medieval times to reuse parchment, so one day, when a monk named Johannes Myronas needed fresh writing materials, he recycled this papyrus, scraping the surface to remove the old ink and copying his prayers over the remains of the original text. No one knows what happened to the prayer book after that, but it ended up in the possession of a Danish philologist named John Ludwig Heiberg in 1908 while he was visiting that Constantinople library. Heiberg was the first to study the not-quite-erased text under a microscope in hopes of deciphering it. He finished an initial (incomplete) transcrip-

---

* Spinach turned out to be the key to unlocking the mystery. Uwe Bergmann, a Stanford physicist at the Synchrotron Radiation Laboratory, heard about the Archimedes palimpsest at a conference in Germany and realized his method for studying photosynthesis in spinach could also be applied to the parchment, without damaging the manuscript. Spinach contains iron; and the ink used on the palimpsest also contained iron, so the same technique could be used.

tion, at which point the book disappeared again, until it was uncovered in that dank French closet ninety years later.

To fully appreciate the significance of this discovery, one needs a bit of historical context. It all started with the problem of curves. In the beginning, there was Euclid, whose geometrically simple world consisted of tidy planes, clean straight lines, and points, with an arc or a circle tossed in now and then, just to add a bit of variety to the mix. It was all lovingly laid out in a thirteen-volume treatise called the *Elements*, perhaps the most influential mathematical text of all time.* Conspicuously lacking in Euclid's collection of geometric postulates and axioms were curves. He was certainly familiar with curvy shapes. There is evidence in the writings of a later Greek mathematician, Apollonius of Perga (circa 160 B.C.) that Euclid studied cross sections of cones, although any treatise he may have written on the subject has not survived.

Cones slice and dice into four basic curves: the circle, the ellipse, the parabola, and the hyperbola. What curve you get depends on the angle of the plane as you're slicing. A simple horizontal slice, for instance, will yield a circle. Tip the plane very slightly, and that circle becomes an ellipse. Tilt the plane such that it runs parallel to one side of the cone, and that ellipse becomes a parabola. Finally, if you tilt the plane so that it intersects the second cone, you end up with a hyperbolic curve.

There are many other kinds of curves, of varying complex-

---

* Abraham Lincoln kept a copy of Euclid in his saddlebag, and studied it late at night by lamplight. "You never can make a lawyer if you do not understand what demonstrate means; and I left my situation in Springfield, went home to my father's house, and stayed there till I could give any proposition in the six books of Euclid at sight," he later wrote.

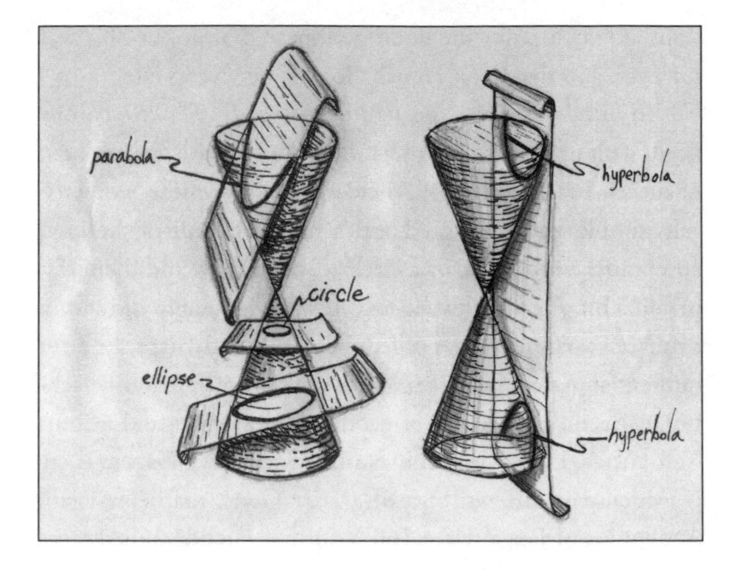

ity, but they are much more than mere geometric curiosities. Curves are geometry in motion. A pendulum swinging back and forth forms the arc of a circle. A falling apple forms a parabola, as does the trajectory of a baseball. The planets move about the sun in elliptical orbits, while many comets move in hyperbolic orbits. A spring bouncing in the absence of friction forms a periodic sine wave. All these types of motion can be represented graphically by a smooth, continuous curve, and that curve in turn can be used to make predictions about the trajectory (path) of a moving object. That makes geometric curves central to the realm of calculus, so naturally that's where I started in my mathematical quest.

DANGEROUS CURVES

Archimedes liked to invent things, and he had a particular knack for fashioning ingenious weapons of war. Legend has it that he built an array of gigantic mirrors of bronze or glass and arranged them in such a way that they were capable of collecting, focusing, and redirecting the sun's rays in order to set enemy ships on fire from a distance—a scaled-up version of frying bugs by focusing sunlight onto them with a magnifying glass. Ancient historical accounts report that he turned this ingenious "death ray" device onto invading Roman ships during the siege of Syracuse in 213 B.C., reducing them to cinders.*

The best shape for an array of mirrors in order to accomplish this is a parabola. Being a thorough sort of inventor, Archimedes set about figuring out how to determine the area under that particular curve. Curves were knotty problems in Euclidean geometry. It's a simple enough matter to determine the area of a triangle or a rectangle, with their clean, straight lines, but ancient mathematicians wrestled mightily with a means for determining the area under a curve. One hundred years or so before Archimedes, a Greek astronomer and mathematician named Eudoxus of Cnidus figured out that while one couldn't calculate the area under a curve exactly, it was

* From an account by John Zonaras, who wrote in the twelfth century A.D.: "At last in an incredible manner he burned up the whole Roman fleet. For by tilting a kind of mirror toward the sun he concentrated the sun's beam upon it; and owing to the thickness and smoothness of the mirror he ignited the air from this beam and kindled a great flame, the whole of which he directed upon the ships that lay at anchor in the path of the fire, until he consumed them all."

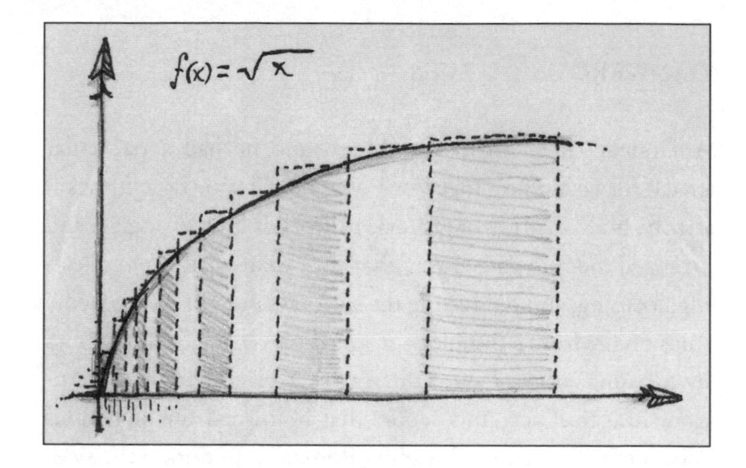

possible to approximate that area by filling it in with a succession of rectangles (see above).

Now you just need to figure out the area for each of those rectangles by multiplying the width by the height. Then add them up, and the result is a rough estimate of the area under the curve. That is known as a first approximation. We refine our calculation by filling in that same area under the curve with a succession of smaller rectangles, determining their areas, and adding those values together. The smaller the rectangles, the more will be needed to fill in that space, and the closer we will get to the actual area. So we do a third iteration, then a fourth, our rectangles getting smaller and smaller each time, until we collapse in fatigue. Eudoxus called this the "method of exhaustion," and never was a term more apt. Very little is known about Eudoxus, but he studied under Plato in his youth, walking seven miles each way from his home in Piraeus to attend Plato's lectures, so he was accustomed to exhaustion.

Archimedes found himself adapting Eudoxus's method to

design his ship-incinerating death ray. He knew quite well how to determine the area of a triangle. So Archimedes drew a triangle inside the parabola, leaving two small gaps. He then drew two smaller triangles inside those gaps, leaving four even smaller gaps, and continued to draw ever-smaller triangles to fill the ever-smaller gaps. Then he calculated the area of each triangle and added them together to get an approximation of the area under the parabola.

Even though the end result was an approximation, it was close enough to the precise area to suit the Greek inventor's needs. More important, it was a critical first step toward defining calculus. With each iteration in the method of exhaustion, the triangles become smaller and smaller, and thus it takes more and more of them to fill the area under the curve. When the number of triangles (or rectangles, in Eudoxus's original method) becomes infinite, that is the point where we get the *exact* area under the curve. And that process of summing up an infinite series of things is the essence of integral calculus.

Archimedes actually may have been less successful at building a viable death ray than he was at estimating the area under a curve. Numerous attempts have been made over the years to re-create this supposedly pivotal moment in the siege of Syracuse, most recently in 2005 by a team of engineering students at MIT. The team built an enormous bronze and glass reflector on the edge of San Francisco bay and tried to focus sunlight onto a small fishing boat about 150 feet away, in hopes of setting it on fire. This didn't work. So the MIT engineers moved the boat closer, to around 75 feet. This time they managed to create a small fire, although it quickly fizzled out.

Part of the problem was cloud cover; the mirrors only work when the sun is shining. Since Syracuse faced east toward the ocean, Archimedes' device would have only been useful in the morning. Then there is the time factor: the death ray did not work quickly. Shooting flaming arrows at the Roman ships anchored in the harbor would have been far more practical and efficient. That was the conclusion of TV's *Mythbusters*, who issued the challenge to MIT after failing in their own attempt to re-create the boat-burning. Executive producer Peter Rees told the *Guardian* that the tale of Archimedes' death ray is mostly likely a myth: "We're not saying it can't be done. We're just saying it's extremely impractical as a weapon of war."

Even if the weapon proved impractical, the exercise of creating it gave Archimedes some valuable insights into geometric curves. His array of flat mirrors formed a makeshift parabola out of straight lines; together they approximated a parabolic curve. Magnify any curved line sufficiently, and it looks more and more like a straight line with each level of magnification. Archimedes realized he could view a circle, for example, dynamically as an accumulation of an infinite number of smaller pieces added together—triangles, again, in this case—rather than as a static, unchanging whole.

This is the method he used to prove how to find the area of a circle: half the product of its radius and its circumference. It's now a standard maxim in geometry textbooks. It worked out so well that Archimedes later adapted Eudoxus's method to calculate the volume of a sphere (a three-dimensional circle) by enclosing it in a cylinder. He considered this solution his greatest achievement, even asking that his tomb be adorned with a sphere contained in a cylinder. Historical records indicate that Cicero, while visiting Syracuse in 75 B.C., located the tomb of Archimedes, which did indeed feature a sphere inside a cylinder.

The problem with the method of exhaustion is that the process literally could go on forever. One would never be able to calculate the *exact* area under a given curve, because how can one draw an infinite number of rectangles or triangles? Managing infinity is a crucial achievement of calculus. The ancient Greeks had an imperfect understanding of the concept of infinity, as do most of us encountering calculus for the first time. It's not something easily grasped by our finite human minds. So Archimedes' methodology still fell short of actually *inventing* integral calculus. Perhaps he might have done so, had he not run afoul of that hotheaded Roman soldier. "Killing Archimedes was one of the biggest Roman contributions to mathematics," Charles Seife drily observes in *Zero: The Biography of a Dangerous Idea*. "The Roman era lasted for about seven centuries. In all that time, there were no significant mathematical developments."

PICTURE THIS

While European mathematics languished in the medieval wilderness, a veritable renaissance was brewing in the East—specifically the rise of Baghdad as a cultural mecca for science and mathematics in the ninth century. The driving force behind this intellectual rebirth was the caliph Hārūn ar-Rashīd, who ruled the Islamic Empire from 786 to 809. He insisted on translating the greatest ancient works on math and science from around the world into Arabic—not just the work of the ancient Greeks, but also the achievements of scholars in India, South Asia, and China. His successor, Abu Jafar al-Ma'mūn, went one step further and established the House of Wisdom (Bayt al-Hikma), a scholarly "think tank" to bring together the Islamic world's greatest minds.

One of those minds belonged to Abu Jafar al-Kwarizmi, whom we can blame for the development of modern algebra. He dreamed up how to use an equation to describe an unknown, the original *x* factor. He's the guy who invented that tedious exercise of "balancing" both sides of an equation by adding, subtracting, or dividing by the same amount on both sides, a plague for high school students to this day. He called his brainchild "comparing and restoring." Since the Arabic word for "restoring" is *al-jabr*, today we know this discipline as algebra.

Al-Kwarizmi did this without the benefit of one little character, literally, that we've come to take for granted. The equal sign didn't exist until the sixteenth century.* He didn't use modern

* A Welsh mathematician named Robert Recorde is credited with inventing the equal sign. He used it first in his 1557 treatise *The Whetstone of Witte*, which introduced algebra to England.

algebraic notation, either. Instead, he expressed his unknowns in words rather than variables, and his equations in sentences. In essence, that is what a mathematical equation is: a sentence reduced to a symbolic shorthand so that the quantities can be more easily manipulated. Algebra is about symbols, while geometry is about shapes, yet they share a mathematical connection, even though it would take another several hundred years after al-Kwarizmi's work before East and West merged. Two French mathematicians, Pierre de Fermat and René Descartes, definitively proved the geometry-algebra connection in the early seventeenth century, thereby forging a crucial link in the development of calculus.

The son of a leather merchant, Fermat was a lawyer by profession, working as a counselor to Parliament in Toulouse. He rose quickly through the ranks, aided by the high death rate of that era, when outbreaks of the plague swept through the city frequently. Fermat himself contracted the plague at one point, but proved to be one of the lucky few to survive. Eventually he became a judge near Toulouse, at a time when heretical priests were routinely burned at the stake. I'd surmise that the intellectually minded Fermat appreciated the fact that judges were discouraged from social interactions, lest they be swayed in their verdicts by conflicts of interest. This freed him to spend most evenings holed up in his study, poring over mathematical proofs to his heart's content.

Sometime in the 1620s, Fermat first encountered a work of Apollonius of Perga called *Plane Loci*, exploring two-dimensional curves. Fermat set about proving (in the rigorous, mathematical sense) some of his ancient colleague's results. He discovered that geometric "statements" of the ancient Greeks could also be rendered algebraically—essentially translating them into *x*'s, *y*'s, and the other accoutrements of symbolic equations.

Any geometric object—a square, a triangle, a curved line—can be represented by an equation. These are the formulae we all had to memorize in geometry class: a circle is $y^2 + x^2 = a^2$, for example, while Archimedes' killer parabola is $y = ax^2$. Points on a graph are noted as sets of numbers inside parentheses (x, y) representing a point in space. The x indicates how far along the horizontal axis a point is located from a point of origin (0). The y does the same on the vertical axis. If you generate enough points from the equation and connect the dots, you end up with a curve. The more points you plot on your Cartesian grid, the smoother the resulting curve will be.

We know these as Cartesian coordinates because the introverted Fermat procrastinated on polishing his work into a publishable format; his ideas didn't appear until 1637, with the publication of *Introduction to Plane and Solid Loci*. That same year, Descartes covered much of the same ground in a separate treatise entitled simply, *Geometry*. Born in 1596, Descartes lost his mother to tuberculosis when he was barely one year old. His father, a member of his provincial parliament, trusted his son's education in philosophy and mathematics to the Jesuit priests at a college in La Flèche. But after earning a degree in law in 1616, Descartes "abandoned the study of letters," opting instead to travel the world to gain as varied experience as possible.

Yet he retained an interest in philosophy and mathematics, and actively pursued knowledge in both. One day, the story goes, he lay on his bed watching a fly buzzing through the air. Descartes realized that its position at any moment could be described by three numbers representing its distance along each of three intersecting, mutually perpendicular axes (corresponding to the lines formed by the intersection of the room's walls in a corner). This insight formed the basis of the Cartesian coordinate system. Descartes—along with Fermat—used this coordinate system to turn figures and shapes into equations and numbers.

While both Fermat and Descartes independently conceived of the underlying notion of translating between curves and algebraic expressions, people liked Descartes' treatise a bit better, mostly because his notation was easier to use. But Fermat is the one who realized that it worked both ways: He could also turn an equation into a graph, and work with the resulting curve to glean insights that might not be readily apparent from simply studying the abstract algebra.

Most notably, Fermat realized that converting the expression into geometry made it easier to find the largest and smallest value within a given range—the maximum and minimum, as we call them today. At any point on a curve, it is possible to draw a straight line that just touches it at exactly that point, called the tangent. You simply study the line that is tangent to the curve at the point of interest and determine its slope.

If the slope of that tangent line is positive (slanting upward from left to right), the expression is increasing; if negative (slanting downward), the expression is decreasing. The steeper the slope, the faster the expression is increasing or decreasing. Where are the maxima and minima? Wherever the slope of the tangent line flattens out to zero (becomes horizontal) along that curve. Like Archimedes before him, who stopped just short of inventing integral calculus, Fermat came within a hair's breadth of inventing differential calculus.

So the area under a curve corresponds to the integral, while

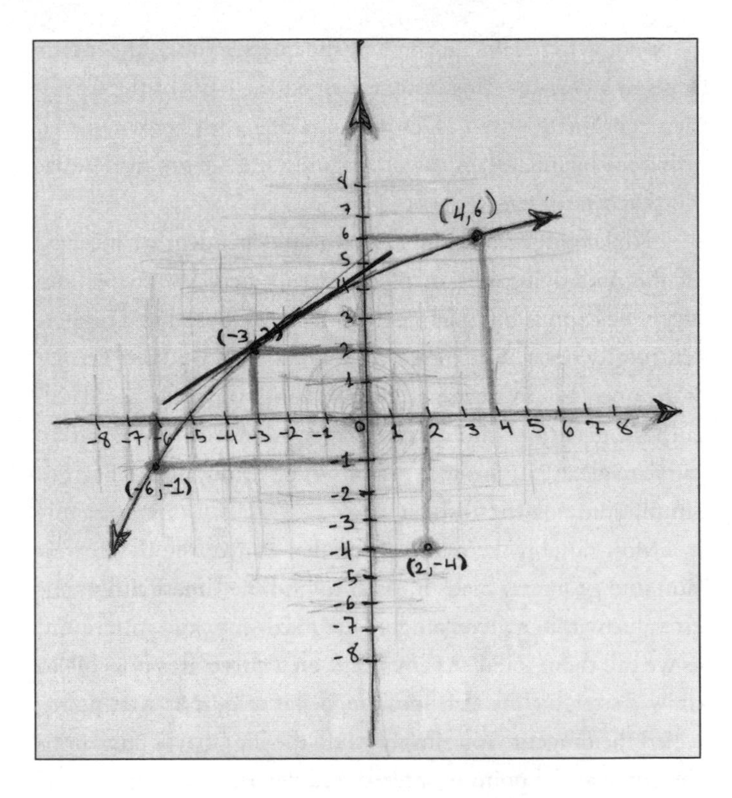

the slope of the tangent line to a point on that curve corresponds to the derivative. With the merging of algebra and geometry, the stage was set for calculus to make its grand entrance. Ultimately, the credit for inventing calculus is given jointly to Isaac Newton and Gottfried Wilhelm Leibniz, who independently made their revolutionary discoveries in the 1660s and 1670s, giving rise to an epic intellectual battle for the title of Inventor of Calculus.

CLASH OF THE TITANS

Isaac Newton hardly needs an introduction. He is almost universally recognized as the father of modern physics via his masterpiece, the *Principia*, as well as his work on the nature of light published in *Opticks* toward the end of his illustrious career. The *Principia* is inarguably one of the most influential scientific books ever written—eighteenth-century mathematician Joseph-Louis Lagrange declared it "the greatest production of a human mind"—yet it is one of the least read. Three volumes of mathematical theory on the nature of gravity and the laws of motion, rendered in excruciatingly pedantic seventeenth-century Latin prose and chock-full of equations, are hardly summer beach reading. Apparently Newton made it deliberately difficult "to avoid being baited by little smatterers in mathematics." The Great Newton despised dilettantes.

The son of a yeoman farmer in Lincolnshire, England, who could neither read nor write, Newton was born in 1642, two months after the death of his father, and so premature and small that hardly anyone expected him to survive. His mother, Hannah, married a clergyman named Barnabas Smith when Isaac was only three years old and promptly moved away with her new husband to start a new family, leaving young Isaac behind with his grandparents.

Hannah wanted him to become a farmer, and when the boy was seventeen, he was expected to take over the family farm. But he proved disastrous at minding the sheep or cows, feeding the chickens, or taking produce to market. Invariably he would be found sprawled under a shady tree with a book, jotting his thoughts down in a notebook, or jumping from one spot to another in the field, trying to determine the length of those jumps. He invented methods for producing chalk and gold ink, and a technique "to make birds drunk," as well as a phonetic alphabet; he "contrived water wheels and dams" and dabbled in magic tricks. In short, he did anything *but* the various chores a competent farmer must master.

Hannah relented and packed Newton off to Cambridge University to pursue the life of the mind, where he earned his undergraduate degree in science and math in 1665. His graduate studies were interrupted by the outbreak of the plague in Cambridge. Students and professors alike fled the city, and Newton returned home for the ensuing year, until the panic (and danger) had passed. He later described this period as "the prime of my age for invention and minded mathematics and [natural] philosophy more than at any time since." He wasn't exaggerating. Not only did he work out his three laws of motion and a universal theory of gravity; he also invented the mathematical tool he needed to achieve those insights: calculus.

"Rather than thinking of a curve as a simple geometrical shape or construction on paper, Newton began to think of curves in real life—not as static structures like buildings or windmills, but as dynamic motions with variable quantities," Jason Bardi writes in *The Calculus Wars*. Take that famous (and possibly apocryphal) anecdote about Newton observing an apple falling from a tree and coming up with his critical insights into gravity. The position and speed of the apple are changing at every moment*: The apple is still on the tree at

* The acceleration is constant once the apple starts falling.

what physicists call time zero. (That's shorthand for "the value of the variable *t* for time is 0.") A fraction of a second later, it has started its fall, and another fraction of a second finds it midway from branch to ground, and so forth. The apple's descent progresses in tiny increments (then called infinitesimals) until it hits the ground or Newton's head. Plot each tiny point describing position versus time along a Cartesian grid, connect the dots, and you end up with one half of a parabolic curve.

Once he plotted a curve, Newton drew on Fermat's prior work and figured out how to find the slope of the tangent line for any point along that curve—the derivative, which he called the fluxion. Then he realized that finding the area under the curve (the integral) represented the process in reverse. Newton's key insight was the connection between the derivative and integral. Finding the area under a curve (integration) is the reverse of finding the slope of a tangent line (differentiation). That is the fundamental theorem of calculus.

Newton noticed other intriguing connections: The apple's velocity is the derivative of its position, while its acceleration is the derivative of its velocity. This also works for the integral. Add up the accumulated rate of acceleration over time, and you get the apple's velocity; add up the accumulated velocity over time, and you get the apple's position. Thanks to the fundamental theorem of calculus, it is possible to change one problem into another problem. If we have an equation that tells us the position of a falling apple, from that we can deduce the equation for the velocity of the apple at any given moment of its fall.

What made Newton's method so revolutionary was its universality: The same equations that can be applied to the speed and position of a falling apple are also applicable to the planets orbiting the sun, the rate at which a cup of coffee cools, how interest accumulates in a savings account—any system in which one quantity is changing with respect to another. So calculus is a nimble beast, a flexible tool that, with lots of practice and a bit of creativity, can take you from a situation where you only have a little bit of information, to one where you have deduced *a lot* more information.

In modern calculus, these quantities—position, velocity, acceleration, and so forth—are known as *functions*, a concept that didn't exist in Newton's time. Here's the kind of textbook definition that, while technically correct, conveys very little actual meaning to the beginning calculus student: "A function is a set of ordered pairs where, for every value of x, there is only one corresponding value for y." But another way to think of the function is as a link between cause and effect. The variables x and y, for instance, are wholly interdependent, such that, if a change occurs in one of them (cause, or the independent variable), the other changes in response (effect, or the dependent variable). Calculus describes this rate of change. In economics, price is a *function* of market supply and demand, rising and falling with the whims of consumer appetites. In physics, potential energy is a *function* of height: The apple's potential energy is dependent on how high it is in the tree's branches, and as the apple falls, that potential energy is converted into kinetic energy.

In the case of Newton's apple, the position function is the entire collection of points that, taken together, describe the apple's position at every single instant during its fall. A similar set of points plotted out for the apple's velocity at any given moment in time comprises the velocity function. But a function is far more than the sum of its parts: It transcends them.

Functions are powerful tools because they confer the power of prediction. You no longer need to perform a new calculation to determine the position or velocity of that apple at each moment in time. With the function, you *know* the apple's position or velocity at every possible moment in time.

Historians generally agree that Newton was the first to state the fundamental theorem of calculus and was also the first to apply derivatives and integrals in a single work (although he didn't use those terms). The problem is that like Fermat, he suffered from publication procrastination. Fermat's dilly-dallying left the field wide open for Descartes to sweep in and claim shared credit for linking algebra and geometry. Newton didn't publish any of his work on calculus until 1704, in an essay entitled "On the Quadrature of Curves" in the back of *Opticks*—quadratures being a fancy name for the areas under curves. By that time, Gottfried von Leibniz's version of calculus was already causing a stir in Western Europe. While Fermat and Descartes had a few testy exchanges, on the whole they maintained an air of civility in their mathematical debates. In contrast, Newton's procrastination led to one of the most bitter controversies in scientific history, dubbed the calculus wars.

Leibniz was born in Germany in 1646, and he was a stellar student even as a very young child. "Precocious" could have been his middle name (in reality, it was Wilhelm). His father died when he was six, so Leibniz was raised by his mother, who encouraged her son's intellectual bent. By eight, he was working his way through his father's substantial library, teaching himself Latin and Greek so he could read the great works of Aristotle and other philosophers. He entered the University of Leipzig at age fifteen and left two years later with his degree in law. Conspicuously absent from his formal education was any study of mathematics; he was entirely self-taught in that discipline.

A chance meeting with the Dutch scientist Christian Huygens ignited Leibniz's interest in the study of geometry and the mathematics of motion; he described their meeting as "opening a whole new world" to him. He pursued these interests in his spare time, inventing in 1671 a handy little machine called the step reckoner. A forerunner of the modern calculator, the device could add, subtract, multiply, divide, and even extract square roots. His reasoning: "It is unworthy of excellent men to lose hours like slaves in the labor of calculation, which could be safely relegated to anyone else if machines were used." Why waste perfectly good brainpower on lowly arithmetic?

At Huygens's urging, Leibniz read Blaise Pascal's work on infinitesimals, as well as the work of René François de Sluse, who had made a rule for constructing tangents to a point on a curve. Leibniz realized that Pascal's approach to infinitesimals could be combined with Sluse's tangent rule and applied to any geometric curve. That same critical insight—the universality of the method—led him to create his own version of calculus independently of Newton.

Leibniz published his first account of differential calculus in 1684, followed by a discussion of integral calculus two years later. It caused a sensation, which rankled Newton's pride; he became convinced that Leibniz had stolen his ideas from his earlier unpublished papers that had been circulating privately in academic circles over the years. (He used his new techniques in his scientific work long before the publication of *Opticks*.) There were rumblings of impending conflict in the ensuing years, as tensions brewed between those in Camp Newton and

Camp Leibniz, but things didn't erupt into outright war until Newton published his essay in *Opticks*.

The opening volley in the calculus wars was an anonymous review of "On the Quadrature of Curves" that appeared in a European journal early in 1704, implying that Newton had "borrowed" his ideas from Leibniz. While Leibniz denied it for the rest of his life, historians generally accept that he was the author. He also engaged in a form of "sock puppetry": He penned numerous anonymous attacks on his archrival's work and then reviewed those attacks (one assumes favorably) in his own signed papers. At the time, Newton was by far the more famous scientist, and a prominent member of the Royal Society of England. While he didn't engage in sock puppetry, he wasn't above using his considerable influence to crush the scientific competition. In addition to Leibniz, during his long scientific career he fought with John Flamsteed, with Huygens, and with Robert Hooke, and each proved to be an acrimonious battle. Newton was not a people person; no wonder he purportedly died a virgin.

In one letter to Leibniz, Newton offered his "proof" that he had invented calculus—but he couched it in a sort of anagram of a Latin sentence. He took all the individual letters and put them in alphabetical order: six *a*'s, two *c*'s, one *d*, thirteen *e*'s, two *f*'s, and so forth. To Newton, it was perfectly obvious: Anyone could simply rearrange all of the letters and find the proof they sought that he, Isaac Newton, had prior knowledge of the key concepts. Very few people felt inclined to go to all that trouble, and frankly, even decoded, the "proof" wasn't especially clear. Roughly translated, the sentence read, "Having any given equation involving never so many flowing quantities, to find the fluxions, and vice versa." That was his stab at summarizing derivatives; Newton would have been a lousy math teacher.

The Royal Society of England sided with Newton on the controversy, crediting him in 1715 with the discovery of calculus. Leibniz wasn't given shared credit until after his death a year later. Today, the consensus seems to be that the two men represent two complementary approaches to the discipline they co-invented. Leibniz was the more abstract of the two, and it's his system of notation that modern scientists still use today, while Newton focused on the more practical applications of calculus. Leibniz can also claim credit for coining the word *calculus*, named for a type of stone once used for counting purposes by the Romans.

Calculus did not find immediate acceptance within the scientific community; there was one final missing piece. The method worked, in that it gave the right answer, but mathematicians found the notion of the infinitesimal deeply troubling. Once again, the problem of infinity raised its ugly head. For instance, Newton relied on a bit of magical hand-waving to make his method work: He argued that since his fluxion units were so small—infinitely close to zero but not exactly equal to zero—they could be ignored for all practical purposes. In his equations, they effectively vanish for no reason. A rigorous explanation for what happens to those fluxion units when an equation is solved would not be found for another hundred years.

Leibniz adopted a symbolic notation—Δx, which stands for a tiny increment—that preserved the infinitesimals yet still enabled mathematicians to manipulate them as if they were actual numbers. (In modern notation, scientists often use *dx* to represent an infinitesimal.) Yet this approach seemed to many mathematicians to be a bit of a cheat. Chief among the naysayers was an Irish bishop named George Berkeley, who in 1734 (seven years after Newton's death) criticized Newton and Leibniz for their fudging of the method, calling infinitesimals "ghosts of departed quantities" and observing that if they were comfortable with that sort of thing, they "need not, methinks, be squeamish about any point in divinity."

TAKE IT TO THE LIMIT

A fictionalized Albert Einstein (portrayed by the late Walter Matthau) plays mischievous matchmaker between his egghead niece, Catherine Boyd, and a good-hearted auto mechanic named Ed Walters in the charming 1994 romantic comedy *I.Q.* Some might object to the considerable liberties taken with historical fact and illustrious personages, but there's a lot to admire in the film, if for no other reason than its inclusion of Einstein's real-life cronies, Kurt Gödel, Boris Podolsky, and Nathan Liebknecht, as supporting characters. As Ed introduces Einstein to Frank, one of his co-workers at the garage, he declares, "This is Albert Einstein, the smartest man in the world!" Intones Frank in his best Joisey accent, "Hey, how they hangin'?"

There is a lovely scene in a diner, where Catherine tries to explain to Ed the gist of one of Zeno's paradoxes. Zeno was a Greek philosopher living in the fifth century B.C. who thought a great deal about motion. Specifically, he speculated that all motion is illusory, and came up with a famous set of arguments to "prove" it. Catherine explains it thus: If she takes one step forward, and then halves the distance traveled with her next step, then halves it again, and so forth, such that the progression goes on for infinity, she will never be able to reach Ed. The distance between them will get smaller and smaller but will never reach zero. The subtext here is Catherine's belief that there is no way to bridge the gap between the couple's intellectual differences and social status. But the practical-minded Ed simply steps over the imaginary line to close the gap: "So how did I do that?" A confused Catherine stammers, "I . . . I don't know." But if she knows her calculus (and she should), the "mystery" should be easy to solve.

Perhaps you've encountered some variation on Zeno's paradoxes before; I certainly had. It pains me to admit this publicly, but I did not realize it was tied to the essence of calculus. In one paradox, Zeno used an arrow flying through the air toward a target*—say, your high school calculus teacher—to illustrate his points, rather than a young couple in a diner, but the basic idea is the same: To reach the target, the arrow must first cover half the distance, then half the remaining distance, and so on, moving an infinite number of times. By that logic, the distance between the arrow and the target would keep getting smaller and smaller, and yet the arrow could never close the gap completely in order to actually reach the target. Your calculus teacher lives to torment you another day.

There's an equally paradoxical corollary: At any given mo-

* Another of Zeno's paradoxes involved Achilles in a footrace with a tortoise. Since Achilles is so much faster, the tortoise gets a head start. Each time Achilles closes the distance by half, the tortoise also moves a bit more ahead. The distance between them gets smaller and smaller, but Achilles can never catch up, since the progression goes on forever. Except in real life, it doesn't, and he can pass the tortoise quite easily.

ment in time, the arrow has a specific fixed position—it can only be in just one place at any given time—which means it is technically at rest (not moving) at that particular instant, even though, taken all together, those individual points add up to an arrow in motion. Motion, after all, is basically the measure of how an object's position has changed over time. But break down motion into infinitely small increments—similar to the individual frames in a film reel—and you find yourself trying to determine how far it traveled in zero amount of time: instantaneous motion. Ergo, the paradox.

In the real world, this doesn't happen, Eventually, the arrow will find its mark, and the calculus teacher will curse the limit with his or her last breath. Ed will close the distance with Catherine, and the two will live happily ever after. This makes the argument a little flimsy by the standards of common sense. But Zeno never intended his paradoxes to be taken literally. The Greeks may have lacked strictly mathematical solutions to the problems, but they certainly recognized the need to reconcile the paradoxes.

Mathematically speaking, the problem is this: Zeno's paradoxes rest on the assumption that the progression will go on for infinity and has no ultimate goal, or limit. But in physical reality, there can be some kind of limit even to an infinite series. That endless series can have a finite sum. In the case of the arrow's tip and its target, as the distances between points become smaller, so does the elapsed time, even if speed remains constant.

The problem of infinity stumped the greatest mathematical minds for two millennia. The Greeks lacked the concept of zero and failed to grasp the idea that a finite distance between two points can be divided into an infinite number of pieces in between. For them, the continuous motion of an arrow in flight is divided into an infinite number of discrete steps, and because there must be an infinite number, the Greeks presumed the arrow would continue flying toward its target forever.

Aristotle tried to get around the difficulty by drawing a distinction between what he called the potential infinite and the actual infinite, arguing that the latter didn't exist. It was fine if a line could always be extended—that would be potentially infinite. But an actual infinitely long line? That would be impossible. Archimedes followed Aristotle's lead: He never claimed that the method of exhaustion would result in the exact value for the area of a curved object; this would require an actual infinite number of triangles or rectangles. He simply said one could refine the approximation as much as one liked—a concept he likewise called potential infinity.

Or so historians and mathematicians believed. That is why the rediscovery of the Archimedes palimpsest in the 1990s is so significant. Heiberg had been unable to transcribe the relevant pages in 1908 because they were so badly damaged. Modern analytic methods uncovered that long-hidden text. This time around, the task of transcribing the fully restored text fell to Reviel Netz, a professor of mathematical history at Stanford University. Netz's transcription hints that Archimedes had a far more sophisticated understanding of the infinite than historians have generally credited him with. In particular, the Greek mathematician flirted with the notion of actual infinity while calculating the volume of a fingernail-shaped figure.

This is not the same as fashioning a rigorous mathematical proof to deal with infinity, however. The man who gets the credit for resolving the problem of infinitesimals is an eighteenth-century mathematician named Jean le Rond d'Alembert. D'Alembert's life story is fodder for an Oscar-worthy biopic. He was a foundling who took the name of the church in Paris on whose steps he was found: Saint Jean Baptiste le Rond. He was raised by a glazier but later discovered his birth parents were a general and a noblewoman.

D'Alembert's insight into Zeno's problem of motion seems obvious in retrospect: that arrow shot from a bow is on a journey, and it makes that journey in an infinite number of smaller steps, but its travel does not continue indefinitely—it has a destination, namely, your calculus teacher's heart. That ultimate destination is the limit, even though the arrow makes an infinite number of subjourneys before arriving. All those subjourneys, added together, mean that the arrow hits its mark—and that summing process is integral calculus.

Let's go back to Catherine and Ed to illustrate. Ed moves closer and closer to Catherine, halving the distance between them with each step. We can add those numbers together: $1 + \frac{1}{2} + \frac{1}{4} + \frac{1}{8} + \frac{1}{16}$, and so on, and notice that the sum of this "infinite series" gets closer and closer to exactly 2. We can check this by "taking the limit": if 2 is the limit, then $2-1 = 1$, $1-\frac{1}{2} =$

½, ½–¼ = ¼, and so on into infinity. With each iteration, the result gets closer and closer to 0. Ed still crosses a distance of 2 feet, but he does it in an infinite number of steps.

This gives rise to one of the most common stumbling blocks for the beginning calculus student: the notion that 0.9999 . . . is actually equivalent to 1. My mind, too, balked at accepting this mathematical fact when my spouse, Sean, patiently explained it to me late one night as I was struggling to understand the limit. Intuitively, we think of a fraction as being a finite sum. We have all eaten one-ninth of a pie, after all. But we also learn about irrational numbers—like π, or the so-called golden ratio, ϕ—where the string of numbers in the decimal expansion really does go on forever.

That was my mistake: assuming that 0.999 . . . is like an irrational number, and therefore represents a sequence of numbers that get closer and closer to 1 but never reaches it exactly. Such is not the case; it is a rational number, in which the same decimal number endlessly repeats, and thus can have a finite sum. Eventually the limit of the sequence equals 1, and that means 0.999 . . . has a fixed value, rather than being an infinite progression. It might seem paradoxical, but two very different mathematical expressions can nonetheless represent the same number. Ergo, to a calculus teacher, 0.999999 . . . is just another way of writing 1.*

---

* My former college English professor, Janet, says that her epiphany on the limit came during a lecture on Zeno's paradox of Achilles and the tortoise, using the number .111 . . . —which is equivalent to 1/9, the point where Achilles catches up with the tortoise (i.e., the limit). Janet didn't take the matter on faith. The woman is a rigorous scholar, so she did all those painstaking calculations herself, adding everything up to find that this endless series of repeating decimal places really did converge to 1/9.

One could argue that calculus itself was invented via tiny infinitesimal bits of accrued knowledge that, taken together, added up to a revolutionary new whole. But like the function, calculus is far more than the sum of its parts, making it possible to understand the world around us in dynamic, rather than static terms. "We live in a world of ceaseless growth and decay, with things in fretful motion on the surface of earth, planets wheeling in the sky," mathematician David Berlinski writes in *A Tour of the Calculus*. "Geometry may well describe the skeleton, but the calculus is a living theory and so requires flesh and blood and a dense network of nerves." Life is constantly moving and changing. Life, in short, is curvy.

2

Drive Me Crazy

> We apprehend time only when we have marked motion . . . not only do we measure movement by time, but also time by movement because they define each other.
>
> —ARISTOTLE

A brooding shot of a long, straight desert highway, shimmering slightly in the heat and stretching far into the horizon, opens Ridley Scott's classic 1991 film *Thelma and Louise*. It's become an iconic image, foreshadowing the women's glorious demise as they drive off a cliff in the Grand Canyon in their 1966 Thunderbird convertible, immortalized forever in celluloid history.

The portion of I-15 that runs between Los Angeles and Las Vegas doesn't stretch quite so dramatically into infinity, but after three long hours of driving under a relentless midsummer sun, it's starting to feel like it could go on forever, particularly since traffic has slowed to a crawl. We have road construction to thank for the delay: The state of California is adding a southbound lane just for trucks, and for some reason, this is also slowing down traffic on the northbound side. Thelma and Louise would have just floored it and blasted their way out, but we are wimpy, law-abiding citizens, and meekly accept our fate.

We're on I-15 in our shiny red Prius because we're hardcore Vegas fans: Taking a weekend jaunt now and then to play some poker, do some shopping, and perhaps indulge in a spot of fine dining or a spa treatment proves quite refreshing. But the road trip also provides an excellent example of calculus in action. Calculus deals with rates of change. Motion is, in essence, change in position with respect to time—however slowly that position is currently changing thanks to the impeded traffic. In fact, at this point, we're barely moving at all, inching along at a scant 10 mph while our stomachs rumble in anticipation of savoring the world's best gyros* and falafel at the Mad Greek Cafe in the tiny town of Baker, California (population 600).

There's precious little to do on a road trip, creeping along a desert highway while breathing in exhaust fumes, with nothing but dusty hills, tumbleweeds, and a long line of rear bumpers as scenery. So I figure it's as good a time as any to muddle through the basics of derivatives and integrals; I'm already bored, hungry, and cranky. Also, it occurs to me that our predicament is reminiscent of Zeno's paradox, outfitted in contemporary garb—exchanging Zeno's trademark toga and sandals for acid-washed Levi's and snazzy ostrich-skin boots, if you will.

Think about it: If our motion is divided into infinitely

* According to the many billboards dotted along I-15 advertising the Mad Greek Cafe.

smaller increments of time and distance—as it would be in a calculus class—in what sense can I claim we are "moving" at all? I can solve this modern paradox by using the tools of calculus to determine our instantaneous speed—how fast we are going at any brief, fixed moment in time—even though our position in time and space is constantly changing. Assuming I know our instantaneous speed (velocity) at every possible moment, can I then use that information to determine how far we've traveled—our position—without cheating and looking at our trusty odometer? Calculus says I can.

ROAD TO NOWHERE

Let's start with a bit of precalculus to demonstrate the concept of instantaneous speed, using the simplest possible example with highly idealized conditions. In my mind's eye, I-15 magically morphs into that endless, perfectly straight road in *Thelma and Louise*, except rather than stretching into eternity, it runs between our home in Los Angeles and the Luxor Hotel in Las Vegas, with an infinite number of points in between.

Imagine that a squad car pulls up as we drive into the Luxor entrance. The officers claim Sean ran a red light a few miles away. Sean denies it. As proof, they show us a time-stamped photograph taken by a traffic camera, showing the Prius just before its nose passes through the intersection. Fair enough, says Sean, but all that proves is that the Prius was at that particular point at that particular time. It merely shows our position, not our velocity. How can they prove that the car was actually moving at that point, rather than stopped at the light? He is a scientist. He demands evidence. He also doesn't want to pay the imaginary fine. To prove he ran the red light, he insists, the officers need to offer compelling proof of the car's instantaneous speed at the moment that photograph was taken.

The officers don't have a radar gun, which measures velocity directly, but unfortunately for Sean, they are well versed in math. They *do* have a time-stamped photograph of the Prius at a similar intersection one minute before. So it's a simple matter for the officers to show where we were at the traffic light—the two-minute mark—and subtract our position at the previous intersection (the one-minute mark) to determine how far the Prius traveled in that time: in this case, one mile. Then they can divide that by the time it took to travel that one mile, and this gives them the car's average speed: one mile per minute, or 60 mph.

Ah, but Sean doesn't give up so easily; he has one more argument to make. The officers are assuming the Prius was moving at a constant speed. Yet every experienced driver knows that one's speed is rarely constant. Just because our average speed was 1 mile per minute doesn't necessarily mean that was our instantaneous speed at the moment we crossed the intersection.

The officers remain undaunted. They don't have access to the information recorded by our trusty speedometer and odometer, so they have supplemented this imaginary stretch of road with some pretty cutting-edge technology, dividing it into intervals at every possible distance and placing tiny nanosize traffic cameras at each and every interval. Call it willing suspension of disbelief, although at the rate nanotechnology research is currently progressing, a scenario quite close to this may one day become a reality.

Thanks to our imaginary nanocameras, the officers have an infinite number of time-stamped still shots of our humble Prius, taken at infinitesimally small intervals along this extremely high-tech futuristic road. This is incontrovertible ocular proof* of the car's position at every given point in time since we left home: In calculus terminology, this is our position function. We know the position of the Prius as a function of time. The cameras reveal that there was *less* time between equal intervals as the Prius approached the light—which means we were actually accelerating.

The basic concept is the same whether we're talking about driving down the imaginary highway at a constant rate or about a more complicated real-world scenario in which our speed is constantly changing. Even though the Prius is accelerating, it still has one specific speed at each instant, and I can use the same highly repetitive process of accumulating evidence to prove it, showing where the car was at all times. I run the same calculation outlined above over and over, for ever smaller intervals, to show how fast the car was going at any given moment in time.

This time, there is a crucial difference: Instead of getting the same answer each time—as in the constant-speed scenario—I get slightly different answers each time. But as the intervals get shorter and shorter, those answers get closer to a point of convergence: 2 miles per minute. The answer is never *exactly* 2. But the answers are clearly converging toward a single answer, to a very close approximation. The limit rears its ugly head. I

* *Othello*, act III, scene 3, line 365.

can safely conclude that the car's instantaneous speed at the moment in question must be 2 miles per minute.

Ingenious, isn't it? Hats off to Newton, Leibniz, and untold mathematicians before and after them who repeated the same exact process of calculation, over and over again, until they'd compiled sufficient proof that the derivative formula works. Thanks to their collective effort, we can simplify this incredibly repetitive process by taking the derivative of our known position function, which will give us our velocity function.* Then we can revert to basic algebra: We take the value for the point in time that we're interested in, and we just plug it into that equation. That will give us our speed at that instant. Behold the power of calculus!

None of this, alas, helps Sean avoid an imaginary traffic ticket. He grudgingly admits defeat. The mark of all good scientists is the willingness to abandon a pet hypothesis if the experimental evidence contradicts it—but that doesn't mean they have to be happy about it.

THE SUM OF ALL THINGS

Taking a derivative is pretty straightforward. Finding the integral is trickier. Conceptually, it's just the flip side of the derivative: With the derivative, I can figure out my car's speed based on how its position changes over time. With the integral, I should be able to determine how far we've traveled in the Prius

* You will find a mathematical breakdown of this process in appendix 1.

based only on measurements of its speed at given locations along our high-tech highway.

Thanks to modern technology, I can just use the car's odometer and built-in GPS system to find the answer. But what if the odometer is broken, the computer has malfunctioned, and we find ourselves stranded in the middle of nowhere, with no other cars in sight? These highfalutin hybrids with their onboard computers and hordes of sensors are pretty sensitive, after all.

Assuming our cell phones still work, we can call AAA, but we need to be able to tell them our precise position. There are no obvious landmarks. "Third tumbleweed on the left next to the giant boulder" isn't going to narrow things down sufficiently. We know we haven't passed Baker. Even if you missed the Mad Greek Cafe—despite the fact that it is gaily painted with the colors of the Greek flag and adorned with plaster replicas of naked Greek statues out front—you'd certainly notice Baker's other main attraction: the World's Tallest Thermometer. Baker is located at the 188-mile mark between our Los Angeles loft and the Luxor in Vegas. Let's say that an hour before we got stranded, we stopped for coffee in Barstow, which is at the 110-mile mark. So I know we are somewhere between 110 miles and 188 miles from our home in Los Angeles.

Had our speed been perfectly constant, this would be a simple task, and we would have no need for calculus. Assuming a constant speed of 60 mph, for instance, and knowing that exactly one hour has elapsed since we left home, I can multiply our speed by the time and conclude that we have gone 60 miles. It's probably a pretty good approximation. But that doesn't reflect actual driving conditions; a car's speed is constantly changing, even more so if there are spots of heavy traffic, and if my lead-footed spouse drives faster than 60 mph to make up for lost time whenever traffic clears.

The only concrete information I have about our velocity is from monitoring the speedometer. Fortunately that's all I need to figure out how far we've traveled and thus pinpoint our location for AAA. The speedometer has displayed our speed at every instant along our journey; taken together, this gives me our velocity function. So I should know exactly how fast I was going at any given moment.

How do I take the variation in speed into account? I set boundaries around the correct answer to get a workable range for determining the distance. First, I do a series of calculations based on the slowest (starting) speed—in this case, at the point where we left Barstow—breaking that journey into smaller and smaller increments of time and adding up the pieces to arrive at a close approximation to the total distance traveled. But this will be an underestimate. So I also need to do the same labor-intensive process for the *fastest* speed the car was traveling over our entire one-hour journey. The resulting approximation will be an overestimate of how far we went, but at least I know that the correct distance is somewhere in between those two values. I then go through the same process for different speeds within the minimum and maximum to further narrow the range. The shorter the intervals of time that I choose to employ, the better, because the speed is less likely to vary by much over tiny times and distances.

In a perfect world, I would have the patience of Job and would continue doing this unbelievably repetitive process at smaller and smaller intervals, thereby getting ever finer approximations of the likely distance traveled. The range becomes smaller and smaller, converging toward a single answer without ever reaching it exactly. In this case, the answer converges toward 172 miles, where the highway intersects with (I kid you not) Zzyzx Road. (Memo to road planners: Buy a vowel already.) Now it is a matter of subtracting the 110-mile mark—our last stop in Barstow—from 172. We traveled 62 miles since stopping in Barstow an hour ago.

I don't determine a precise location via any single division of the interval of time; I get the answer via an infinite number of increasingly improved approximations. Although this exercise in precalculus merely gives me a series of approximations, at some point the intervals become so small that the difference between approximations becomes trivial. AAA can probably find us if we tell them we're within five feet and ten feet of the intersection of I-15 and Zzyzx Road. Integral calculus can simplify matters greatly. Fully integrating speed over time using the velocity function would give me an *exact* answer for my position. Think of it as Eudoxus's method without his exhaustion.

DERIVER'S ED

If we can closely approximate our instantaneous speed and position using the precalculus methods outlined above, it's reasonable to ask why we need calculus at all. It all comes down to functions. Rather than performing an infinite series of calculations for every point along a given curve, the function gives us the value for each of those points all at once, saving us considerable effort and time. Functions confer tremendous predictive power. More important, functions are connected to each other in valuable ways: Velocity is the derivative of position, and acceleration is the derivative of velocity. We integrate acceleration over time to find the velocity function, and we integrate velocity over time to find our position function. These connections let us make inferences based on what we do know, to figure out what we don't know.

In *Zero*, Charles Seife compares a standard equation to a machine in which you punch in a number and get another number back. That's what functions do. Plug any number into a function, and it will give you a new number. Taking a derivative or an integral does the same thing, except you feed the machine a function and it sends back a new function. It's just a higher level of abstraction. That is how, using calculus, we can transform one problem into another. "Nature doesn't speak in ordinary equations. It speaks in differential equations, and calculus is the tool you need to pose and solve these differential equations," Seife writes. "Plug in an equation that describes the conditions of the problem . . . and out pops the equation that encodes the answer."

The "plug and chug" method might get you through high school geometry and algebra, but rote memorization of every function, along with its derivative and integral (if known), won't be enough to succeed at calculus. At its core, calculus is about creating and solving logic problems—a most creative endeavor. In fact, constructing a calculus problem is akin to telling a story; we're just doing it with numerical symbols instead of words.

Every narrative has a logical progression, and so does every calculus problem. You identify your central characters and sketch an outline of the plot to create a structural framework. Then you color in the details as you go. The story can be as simple and straightforward as *The Cat in the Hat* or as complicated as James Joyce's *Finnegan's Wake*, but in each case it evolves naturally from the starting point of setting the narrative parameters. Writers and physicists alike spend a great deal of time staring at a blank sheet of paper (or computer screen), waiting for inspiration to strike. This phenomenon can be witnessed firsthand on any given night at our house.

Let's revisit our two idealized scenarios from the perspective of a narrative. Who is my main character? In the first example (Sean attempting to avoid an imaginary traffic ticket), it would be position, because that is the accumulation of data available to us—what we already know. At every point in time, our Prius has a position on the road; all those points taken together comprise the position function (position as a function of time), which we can represent algebraically as $p(t)$, where p stands for position,* and t stands for time. Note that I picked p because it's easy to remember; I could have called it x or q or even *Sally*, and it would still stand for the exact same thing in this context: position. It is the context that gives a particular variable its meaning.

We can graph every single value for p as a point on a Cartesian grid and connect the dots to get a curve. Now we have a "face" for our main character, the position function. That means we can plug different values into this equation to find where we are at any point in time using basic algebra.

Sean admits that usually, collecting data from the real world doesn't give us a simple function, "but as physicists we

* Physicists are probably freaking out reading this, since they habitually use p to denote momentum, having already assigned m to denote mass in their equations. But it's the context that gives the variable meaning, so for now, I'm sticking with p.

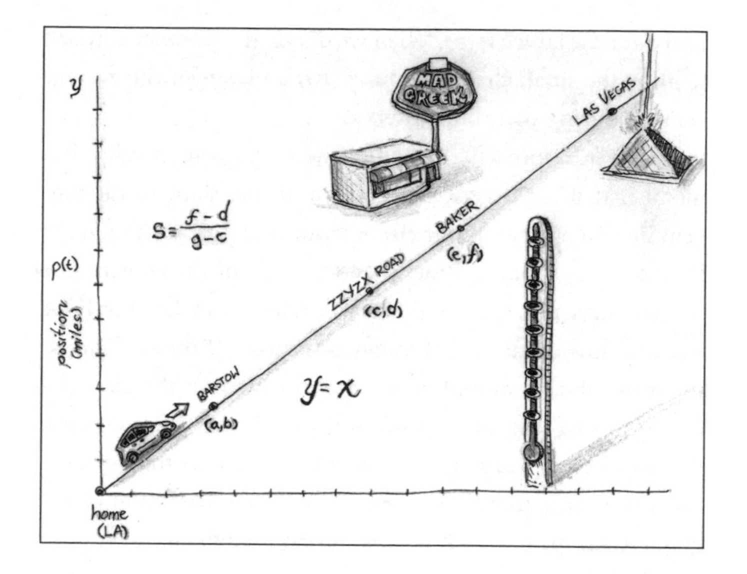

often find it useful to approximate the messy real world by some simple function that we can write down cleanly." Fair enough: Plenty of writers take liberties with narratives, too, if it makes for a better story.

What is the main character's ultimate goal? Given the "clues" about our known position, we want to figure out how fast we are traveling at a particular point on our trip. There is even a central conflict: How does the main character reach that goal? It's a process of deduction, using the clues we've been given: namely, our position function. We can take the derivative of the position function—a process of subtraction and division—to find the corresponding velocity function, which we can use to determine our instantaneous speed at any given point. To do this, we start with our current position (p), take our position a tiny bit into the future, then subtract the two to find out how far we went. Then we divide the distance traveled (Δp) by the small change in time (Δt) and we get the average velocity during that short interval.

We can approach the same question geometrically. Remember that the derivative also gives us the slope of the tangent line on a curve. If our curve represents the position of the Prius at every point in time, then the slope of the tangent line to that curve at a specific point will tell us how fast the Prius was traveling then: the instantaneous speed. If the car is moving forward, that motion will be represented on the graph by a tangent line slanting upward; if the car is moving backward, the tangent line will slope downward. The steeper that line, the faster the car is traveling. The minimum or maximum of the graph has slope 0, which means the car is stopped.

How do you find the exact slope of the tangent line? You draw a straight line between two points on the graph and then look at how much that line rises or falls (the y axis) over that set distance (the x axis) between two points. We get the derivative by looking at ratios—for example, a difference in the position of a moving car at two separate times—so the slope of that line is the fraction of the change in position divided by the change in time. You do the same thing again with two closer points; and so on, until all those straight lines converge to a tangent line whose slope is equivalent to our instantaneous speed. The closer those points are to one another, the closer we can approximate the slope; we have the exact answer when there is no distance between those two points. This is a visualization of the limit: the difference in height goes to zero and so does the distance between the two points.

Now let's revisit the integral via the second example: figuring out how far we have driven based on our velocity. We are telling the same story from another character's point of view, and it changes the "narrative" in some crucial ways. In this case, our main character is the velocity function. We don't know the position; we know the velocity, and we want to deduce our position from that. We've seen that it is possible to figure out how far we've driven knowing just the velocity of the Prius at each instant along the road—the velocity function—using that tediously time-consuming precalculus method. Since we have a "face" for our function, we can determine the area under that curve between the two points of interest via our old friend Eudoxus and the method of exhaustion.

There is a shortcut: I would get exactly the same answer if I simply subtracted our beginning position from our ending position. Of course, I don't *know* our exact ending position, which complicates matters. All I have is the velocity function and my known starting position. My myriad calculus books assure me that all I need to do is figure out which position function generates the known velocity function by taking an integral, then use that position function to determine where we are when our Prius has its hypothetical breakdown.

How do physicists find the integral they need in the real world? They usually look it up. Seriously. A lot of this work has already been done by the generations of mathematicians who came before us, bless their detail-oriented souls, so why waste valuable time recrunching all those numbers? Most standard calculus textbooks contain tables of known functions for both derivatives and integrals to assist beleaguered students—or their teachers provide them with formula sheets. Sean ditched his calculus textbook long ago; instead, he has a big blue book called *Standard Mathematical Tables*, filled with nothing but a bit of explanatory text and lots of incomprehensible notations.

It's now also possible to download calculus apps for your iPhone. The problem is that it is impossible to list every single integral. Even *Standard Mathematical Tables* soberly admits its own shortcomings: "No matter how extensive the integral table, it is a fairly uncommon occurrence to find in the table the exact integral desired."*

Occasional patterns do emerge. For instance, there is a mathy trick we can use to help us find the desired derivatives and integrals for any constant times x. Remember that the derivative and integral are opposite processes: Each undoes the work of the other. The integral is a process of multiplication and addition. If we are given the function $2x$ (2 is the constant, meaning it is unchanging), an integral of $2x$ is the function x^2. Because the derivative is a process of subtraction and division, that means that the derivative of x^2 is $2x$. Similarly, for $2x$, the derivative is the function 2. Indeed, Sean explains that this is pretty much a universal rule.†

I know what you're thinking: *I thought 2 was a constant. How can it also be a function?* That confused me, too, at first. Sean explained that in the above example, 2 plays different roles, depending on the context. It plays a constant in the function $2x$. But then we took a derivative, an operation that gives us a new function back: Now 2 is playing the role of a function. Technically, it's the dependent variable (generically repre-

---

* The eighteenth-century mathematician Johann Bernoulli, whom we will meet in chapter 8, also appreciated the difficulty. "But just as much as it is easy to find the differential (derivative) of a given quantity, so it is difficult to find the integral of a given differential," he once wrote. "Moreover, sometimes we cannot say with certainty whether the integral of a given quantity can be found or not."

† The derivative of ax^N is anx^{N-1} (a times n times x times x^{N-1}) for any constants a and n. Likewise the integral of ax^N is equal to $\frac{ax^{N+1}}{(N+1)}$. Now aren't you sorry you asked?

sented by y). Plug in any random number (x, or the independent variable), and the function will send that number to 2. Think of it as an ordered pair (x, 2), where x can be any random number. The point is, in this particular scenario (a constant times x), whenever we have a derivative formula, we can automatically find an integral formula.

Once we've identified the integral we need, we don't have to resort to the tedious process of dividing up the area under the curve into tiny pieces and multiplying and adding ad nauseam. Instead, we just subtract the value of the integral at the end of the curve from the value at the beginning of the curve to get the answer. Let's say we want to take the integral from 1 to 4 of the function x^4. We can rely on our little trick above to determine that an integral for x^4 is $\frac{x^5}{5}$. Now we simply plug in the highest and lowest values for x in the range of interest (1 to 4) and subtract the results. Our answer: 1,023 divided by 5, or 204.6. This means that we have gone 204.6 miles between those two points—or that the area under the curve between point 1 and point 4 along the x axis is 204.6.

A physicist who blogs anonymously at Gravity and Levity describes physics and calculus at the high school level as a kind of game. "It was like a little logic puzzle where the rules of the game were given to you (usually on a formula sheet) and you were asked to use them cleverly to come up with a solution," he says. "A friend of mine once put it succinctly: 'Physics is all about finding out which variables you know and which variable you want, and then searching through your formula sheet for an equation that has all of those letters in it.' That, more or less, was the physics game. You rearrange some symbols on a paper and you come up with an answer. Instant gratification."

Some students take to the game quite naturally; others, like me, do not. But none of us will realize the full power of calculus until we move beyond treating it as a game and learn how to use it creatively to solve real-world problems.

YOUR MILEAGE MAY VARY

Even if we lose at the poker tables, I've gained something tangible from our weekend excursion: a valuable insight into the fundamentals of calculus. The derivative and integral are two different ways of looking at the same situation, namely, our Prius driving down a straight, level road. I can use the derivative to find our speed from our position and use the integral to figure out how far we've traveled based on our speed.

The speedometer and odometer in the Prius do these sorts of calculations all the time. It was quite ingenious of human beings to build these handy little devices whose primary purpose is to determine the exact information about speed and position that early mathematicians so meticulously calculated by hand. What is their secret? They have much more real-time data at their disposal. Both the speedometer and odometer are designed to collect every possible data point (for speed and distance, respectively) that it can in real time. The speedometer gives us a velocity function; the odometer gives us a position function. We can pretty much find out anything we need to know with this information, with no need to resort to calculus.

Speedometers measure the speed of a car by counting every single rotation of the tires. In older cars, they are mechanical, connected to a drive cable snaking its way from the transmission to the dashboard instrument cluster. The drive cable is basically a cluster of tightly wound coil springs wrapped around a center wire. When the wheels of a car turn, the gears in the transmission turn, and their rotational speed is sent down to the speedometer, where it can be measured and displayed.

The Prius speedometer is electronic (as is the odometer) and gets its rotational data from a vehicle speed sensor mounted to the crankshaft, rather than a drive cable. The sensor is little more than a toothed metal disk and a simple detector covering a coil that emits a magnetic field. The teeth interrupt the magnetic field as the disk rotates past the coil, creating a series of pulses, which are sent to the car's computer via a single wire. The computer counts the magnetic pulses as each tooth of the metal disk passes by the coil. The real-time speed is displayed on the speedometer, so we can keep track of how fast we are traveling. The speedometer is linked to the digital odometer, so for every forty thousand pulses, the odometer adds one mile.

In fact, the Prius onboard computer goes even further: It combines the data on speed and distance with data collected from sensors monitoring gas usage to determine how many miles the car is traveling for each gallon of gas it consumes—both in real time, and on average over a given period. All this information is processed and presented in a colorful digital display that constitutes a real-time video game, showing how much gas you use at any given moment, and how your driving behavior can change that consumption for better or worse. Really, it's a miracle that we Prius drivers manage to avoid plowing into ditches and rear-ending other cars all the time, given how distracting it is to have that constantly changing dynamic information on display before us.

Let the naysayers knock my plucky little hybrid if they must, but thanks to that real-time graphic display, I am now hyperaware of how much energy I consume when driving, and how much even tiny changes in driver behavior, type of terrain, or weather conditions can affect my overall mileage. By virtue of constant feedback on your fuel-efficiency performance, the Prius trains you to be a more energy-conscious driver. For instance, accelerate gradually, and you'll use less energy than if you put pedal to the metal in a vain attempt to go from 0 to 60 in a few seconds.

Also, traveling at a steady speed, even in heavy traffic, is better than jerkily starting and stopping, because every time you restart after a full stop, you have to overcome the car's inertia all over again. I try to leave a bit of extra distance between my car and the vehicle just ahead, so I can coast a little rather than brake suddenly. Under the best conditions, the difference can be as significant as getting 75 miles per gallon versus 25 mpg. I reflect on that whenever I feel frustration at Los Angeles' notoriously congested freeways. I might be inching along at a snail's pace, but I reap the benefit by averaging many more miles per gallon, even if it takes longer to reach my destination. Collectively, these practices have become known as hypermiling.

Even traveling at a steady speed, in general, the faster you go, the more energy it takes to maintain that speed because of increased air resistance (drag). The engine has to work constantly to overcome the resulting drag and thus consumes more fuel. It's tough to correctly calculate the drag coefficient for anything but the simplest of shapes, but in general, at high speeds, the drag force increases as the square of the velocity. In plain English, this means that if you're traveling at 100 mph, you'll experience four times the drag force you'd experience if you were traveling at 50 mph.*

Small increments in improved fuel efficiency can add up significantly over time. So driving just at (or slightly under) the speed limit can result in considerable energy savings in the long term. Back in 1974, the federal government instituted a 55 mph speed limit on highways, not because it was safer† but because it conserved fuel at a time when oil was scarce. Similarly, driving uphill uses more energy than coasting downhill—any avid bicyclist could tell you that—as does driving into a strong headwind. Certain driving conditions are beyond one's control. Don't even get me started on what a ten-hour drive from Salt Lake City to Los Angeles in gusting crosswinds through a mountain pass did to my average miles per gallon.

Why doesn't everyone ditch their current gas-guzzling cars for a Prius or similar hybrid? The answer might surprise you. It turns out that many of us assume that saving gas (and therefore money) corresponds linearly with miles per gallon. But according to a June 20, 2008, article in *Science* by Richard Larrick and Jack Soll at Duke University's Fuqua School of Business, the gas used per mile is actually *inversely proportional* to miles per gallon. They call this the mpg illusion.

---

* Yes, a Prius can get up to those speeds, as we learned in 2007 when former vice president Al Gore's son was pulled over for going 110 mph in his hybrid. And the car's sleek aerodynamic shape means it has a lower drag coefficient than, say, the boxy Scion xB.

† It may very well be safer to drive more slowly, according to a 2008 study by scientists at the University of Adelaide in Australia. They found that the risk of serious injury or death from a car crash doubles for every 5 km/h above 60 km/h. So if you're traveling at 65 km/h, you are twice as likely to be involved in a serious or fatal crash; at 70 km/h, that risk is four times as high. This is because drivers need at least 1.5 seconds to respond to a perceived danger, and the faster one travels, the less time there is to react.

Let's say you own two cars: one with a 34 mpg rating, like Sean's old Toyota Corolla, and another with an 18 mpg rating, like my father's beat-up Chevy pickup. Should you replace the 34 mpg Corolla with a pricey 50 mpg hybrid, or the 18 mpg pickup for a cheaper 28 mpg nonhybrid vehicle, in order to achieve optimal savings? You want to optimize those gas savings to recoup your initial capital investment as quickly as possible. Run the numbers, and it becomes apparent that replacing a 34 mpg car with a hybrid that gets 50 mpg will save you 94.1 gallons of gas per 10,000 miles; in contrast, replacing the 18 mpg truck with a 28 mpg vehicle will save you a whopping 198.4 gallons per 10,000 miles.

That means you're much better off replacing the lower mpg vehicle (the Chevy pickup) with a cheaper alternative to the Prius to get the biggest cost savings. This seems counterintuitive. After all, you're getting a 16 mpg improvement in the first

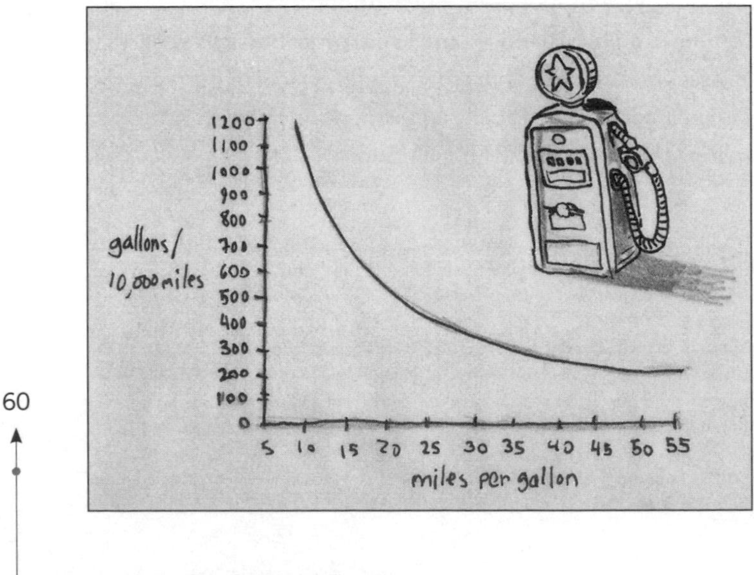

example, and only a 10 mpg improvement in the second. But if you put this data into graph form, you can clearly see the gas used per mile is inversely proportional to miles per gallon.

There is a steeper slope at lower mpg ratings and gradually diminishing returns as one moves up the graph to increasingly higher mpg ratings. So even such seemingly simple numbers can be deceptive, particularly since most of us are sadly deficient in our grasp of basic mathematical concepts. And in this case, our ignorance could prove costly.

That's why I resist the occasional twinge of Prius envy when I read about the 2010 Prius with the solar-powered sunroof and even better mileage. Based on the above calculations, it simply isn't cost-effective to replace my 2007 model with the newer model; it would take much longer to recoup that capital investment. I'm better off just driving my existing Prius into the ground.

MODEL BEHAVIOR

If digital speedometers and odometers do a better job than manual calculations of speed and distance, why do we still need calculus at all? Calculus is a vital part of almost every field of science because it enables scientists to construct mathematical models to study complicated real-world systems—including traffic patterns. Like the computer dashboard displays in the Prius, mathematical models are visual representations of abstract concepts, with the added advantage of enabling scientists to make useful real-world predictions.

Admittedly, not all mathematical modeling has a practical application. Topologists, for example, are interested in studying imaginary multidimensional shapes that simply couldn't exist in our four-dimensional space-time. But much of the appeal of mathematical modeling for less exalted minds lies in how it can help make predictions about how a system is likely to behave, so we can make better, more informed decisions—such as whether to stay on a clogged freeway and wait out the congestion or try to find an alternate route to avoid any more potential slowdowns up the road. (The latter is not an option on I-15. There is no alternate route.)

The more data points you have to work with, the more accurate your models will be. Ideally you would like a continuous stream of real-time data rather than a collection of discrete data points. That's why state and federal agencies spend about $750 million each year on traffic monitoring to gather better data in hopes of building better predictive models of traffic flow. For instance, several state transportation agencies—Maryland, Virginia, Missouri, and Georgia—are experimenting with software that uses radio signals from drivers' cell phones as tracking devices to monitor traffic patterns. The phones just need to be turned on; the agencies swear they are not monitoring actual conversations.

"Listening posts" are placed throughout a designated region; they are capable of detecting but not sending radio signals. A post will pick up a cell-phone signal and time-stamp the signal's arrival. By analyzing how long it takes the radio wave to reach the listening post from the cell phone, a computer can calculate almost precisely where that phone is located on the highway. You need data from three such listening posts to determine a two-dimensional position of a given cell phone user. Adding radio tags along the highways to time when vehicles pass between given points can determine the car's location and speed. Berkeley, California, has a test-bed project dubbed Smart Cars and Smart Roads, whereby participating cars are equipped with wireless technology to pick up signals transmitted from sensors embedded in the road on which they are traveling. In this way, they can relay critical information, such as whether there's been an accident up ahead, and also serve as anonymous data collectors.

Traffic jams are a bit like the process of freezing. On a sparsely populated highway the cars are far apart and can move freely at whatever speed they choose while maneuvering between lanes—much like the movement of molecules in a gas. In heavier traffic, the "car molecules" are more densely packed, with less room to maneuver, so cars move at slower average speeds and traffic behaves more like a liquid. If the car molecules become too densely packed, their speed is reduced and their range of movement is restricted to such an extent, they can crystallize into a solid, akin to that critical temperature/pressure point at which water turns into ice.

It's a useful analogy, but the reality is a bit more complicated. A physicist named Boris Kerner has analyzed data collected from several years of traffic monitored along German highways and found that traffic tends to follow the rules of self-organization. His model breaks down traffic into three basic categories: freely flowing, jammed (a solid state), and a bizarre intermediate state he calls synchronized flow, in which densely packed car molecules move in unison, like members of a marching band. When all the cars are traveling at roughly the same average speed because of the vehicle density on the roadway, they become highly dependent on one another, or "highly correlated."

When cars are highly correlated, a tiny perturbation will

send little ripples of slowdowns through the entire chain of cars behind the offending vehicle. What happens if the law-flouting driver in the Audi ahead of you decides to text his girlfriend and then has to brake too suddenly when he looks up and realizes he's about to rear-end the BMW just ahead? That makes you brake too suddenly, and the person behind you, and so on.

That's one reason why traffic jams are so common at freeway entrance and exit ramps, or—like on the I-15—when lanes are closed due to road construction (or a major accident). A state of steady synchronized flow, punctuated by these tiny ripple effects, can persist indefinitely, but the balance is delicate and highly unstable. If the volume of cars continues to increase, the density also continues to increase, and eventually you get a "pinch effect": that frustrating stop-and-go phenomenon we are experiencing on the road to Vegas, in which you escape one narrow traffic jam only to encounter another a little farther down the road, until they all converge into a single wide jam. Traffic comes to a standstill.

Given world enough and time, even the worst traffic jams eventually unsnarl. We finally break free of the construction zone, and Sean gleefully accelerates to full speed. Zzyzx Road can eat our dust. Soon we're happily chowing down on gyros and falafel at the Mad Greek Cafe, bagging some tasty pistachio baklava for the road, as well as some Alien Beef Jerky from the tiny store decked out in UFO paraphernalia down the street. (Baker's a pretty colorful town.) An hour or so later, our hunger sated, we are cruising down the infamous Las Vegas Strip toward the Luxor Hotel, where Lady Luck—and the calculus of probability—will determine our fortunes in the casinos.

3

Casino Royale

> The theory of probabilities is at bottom nothing but common sense reduced to calculus.
>
> —PIERRE-SIMON DE LAPLACE

Legendary Vegas gambler Nick the Greek (aka Nicholas Andrea Dandolos) won and lost over $500 million in his lifetime by his own estimation, moving from rags to riches and back again countless times. Along the way, he met pretty much everyone, from Al Capone and Bugsy Siegel to the Marx Brothers, Ava Gardner, and John F. Kennedy. Every celebrity who visited Vegas wanted to meet the last of the true high rollers. So when the world's most famous physicist, Albert Einstein, came to town for a symposium, naturally he sought out Nick the Greek, while indulging in a spot of gambling himself at the craps and roulette tables.* Realizing that his gambling cronies

* "Einstein is gambling as if there were no tomorrow," an eminent physicist is said to have remarked. His companion replied, "What troubles me is that he may know something!"

would have no idea who Einstein was, Nick the Greek simply introduced him as "Little Al from Princeton—controls a lot of the action around Jersey."

Sean is tickled when I tell him this (possibly apocryphal) story. Any serious gambler should have a cool nickname, he declares, and promptly dubs himself S-Money for the duration of our stay. We normally focus on poker when in Vegas, but this time, we're interested in learning craps, because it is a natural fit for discussing the calculus of probability. Much of probability theory emerged from attempts to analyze games of chance, particularly those involving the throwing of dice, sticks, or bones. There is even a theorem known as the craps principle, dealing specifically with event probabilities under repeated trials. And what better way to explore that principle than to hit the craps tables in a bona fide casino?

There are many online guides to playing craps, some with in-depth analysis of all the related probabilities, but these tend to be dense and jargon-heavy. There are also online computer craps games where you can practice placing bets and rolling virtual dice without risking any actual money. But sooner or later, you have to step up and put your wallet on the line. Craps doesn't really begin to make sense until you get your hands dirty and play the game in a real-world setting—like Las Vegas.

Craps is a raucous, fast-moving game—there is one roll of the dice every twenty seconds or so—and this pace can be intimidating for the average newbie still struggling to grasp the rules. So we have opted to take one of the daily introductory classes offered by the New York, New York casino. Our instructor is a dapper man, slight of build, with tidy salt-and-pepper hair, wire-rimmed glasses, and a wry sense of humor, whom I dub Dominic. Dominic has been a dealer for thirty years and is happy to share not just the rules of craps but colorful anecdotes from Las Vegas history.*

He starts with basic protocol: how one handles the dice. When a new craps table opens, for instance, the dealer opens a fresh, factory-sealed pack of five dice, from which the inaugural "shooter" must select two. "Whatever you do, don't grab all five dice, toss them across the table, and yell, 'Yahtzee!'" Dominic cautions. Then everyone will *know* you're a rube.

The dice must be held in one hand, to prevent players from surreptitiously switching in loaded dice. You aren't allowed to rub the dice between both hands for the same reason, or kiss the dice ("You don't know where they've been," quips Dominic), and while it's fine to lightly blow on the dice for good luck, we were advised not to get spittle on them, out of courtesy for the next shooter. By order of the Nevada gaming commission, the casino also requires that both dice bounce off the far wall of the table on each roll, lest certain players try to "rig" the roll. We all take turns practicing this. Dominic warns us not to throw the dice *too* hard, but that doesn't stop one overexcited shooter from tossing them so forcefully that they bounce off the table—narrowly missing a drop down a buxom brunette's cleavage.

---

* According to Dominic, the origin of the term *eighty-sixed* dates back to the days when the Mafia ran Vegas casinos. Whenever cheaters were caught, the pit boss would instruct his henchmen to "eighty-six that guy"—code for taking the victim eight miles out of town and burying him six feet under.

THE DUKES OF HAZARD

Some version of craps has been around for centuries, although historical accounts quibble over the details. Did craps derive from an old game called hazard, popular with English knights during the Crusades as they laid siege to a castle called Hazarth in 1125 A.D.? Perhaps the game is Arabic in origin (*al-zar* in Arabic translates as "the dice"). Or does craps reach further back in history to the Roman Empire, when soldiers fashioned rough-hewn dice out of pig knucklebones? There are certainly references to the game in Chaucer's *Canterbury Tales,* and it was hugely popular in France by the seventeenth century, especially among the aristocracy.

We can credit a French-Creole American nobleman with the tongue-twisting moniker of Bernard Xavier Philippe de Marigny de Mandeville for bringing craps to America. The son of a count, Marigny was born to wealth and privilege on the family's New Orleans plantation in 1785, and his upbringing did much to foster a sense of entitlement. Local lore tells of the 1798 visit to the Marigny estate by the Duc d'Orléans, Louis-Philippe (later crowned king of France in 1830), and his two brothers, and the lavish revels that ensued, including the manufacture of special gold dinnerware. In a show of excessively wasteful extravagance, the gold place settings were purportedly tossed into the river when the festivities ended, for no one would be worthy to eat from any plate used by Louis-Philippe. (One hopes the poverty-stricken locals trawled the river bottom and scavenged the discarded loot.)

With such role models before him, it is small wonder that young Master de Marigny matured into a spoiled, dissolute, and extravagant young man, coming into his substantial inheritance at the tender age of fifteen after the death of his doting father. His long-suffering guardian despaired of controlling the headstrong teenager and shipped Marigny off to London in hopes that there he might learn some temperance. Instead, Marigny frequented any number of gambling dens, most notably the infamous Almack's. That's where he learned the game of hazard, bringing a simplified version of it back home to New Orleans a few years later. In local dialect, the game was dubbed *crapaud*, from a derogatory term for the French in New Orleans, Johnny Crapaud. English-speakers later shortened the name to craps.* The game quickly spread to the Mississippi riverboats and beyond.

Bernard de Marigny died penniless in 1868, having repeatedly subdivided his once vast plantation into numerous land parcels, selling them off to cover his ever burgeoning gambling debts. He is largely forgotten, but two legacies remain: the Faubourg Marigny neighborhood of New Orleans (built on the site of the old Marigny estate), and the game of craps, which is more popular today than ever. In fact, there is a street in the Faubourg Marigny district named Craps, reflecting its founder's place in gambling history.

There have been many refinements to the rules of craps over the centuries, but the fundamentals remain unchanged. Each player takes turns being the shooter, rotating around the table as each individual game comes to an end. Every game

---

* *Crapaud* is French for "toad," you see, and the French are oh-so-fond of eating plump, juicy frog legs sautéed in butter and lots of garlic. An alternative theory is that the name is a corruption of a losing throw in hazard, called crabs, but that explanation lacks the jaunty panache of the *crapaud* theory.

starts with a "come-out" roll: Players place their initial bets on the "pass" line (required in order to play), and the shooter rolls the dice. If the shooter rolls a 7 or 11, everyone who placed a pass bet wins outright. If the shooter rolls a 2, 3, or 12, everyone loses outright. If the shooter rolls any other number, that number becomes the "point" for the duration of the game.

Our first shooter is a middle-aged man of Eastern European descent—let's call him Yuri—visiting Vegas with his wife. He starts off strong, rolling a 7 right off the bat, and the table cheers in victory. We collect our winnings, place new line bets, and Yuri rolls again. It's an 8; this becomes the point, and the game's afoot. Now that a point has been established, we keep betting and Yuri keeps rolling until he rolls the point again (an 8) or he rolls a 7 (craps). If the former, we win; if the latter, we lose. Either way, the game ends, a new shooter takes over, and the cycle begins anew.*

If that were all there were to craps, it would become boring very quickly. So as the game evolved, additional types of bets were added, each with its own set of odds. For instance, as an alternative to the standard pass bet on the come-out roll, a player can place a "don't-pass" bet—essentially betting against the shooter and the rest of the table. One caveat: This will make you very unpopular. It's a very social game, and players tend to bond at a craps table, because people's fortunes rise and fall with the shooter's. Betting against the shooter is a buzzkill. For don't-pass bets, the win-lose rules are reversed. If the

* In May 2009, a middle-aged woman from New Jersey named Patricia Demauro set a new record for the longest craps roll in recorded history: four hours and eighteen minutes. It was only her second time playing craps. She finally lost after 154 rolls of the dice.

shooter rolls a 2 or 3, a don't-pass bet will win outright, while the rest of the table loses. If the shooter rolls a 7 or 11, a don't-pass bet will lose outright—and everyone at the table who placed pass bets will revel in Schadenfreude.

The key difference is if the shooter rolls a 12. In that case, a don't-pass bet will neither win nor lose; it would be a "push." This is simply a means of maintaining the house advantage: Three numbers are losers while two are winners on the come-out roll if you place a pass bet. In contrast, two numbers are losers and two are winners if you place a don't-pass bet on the come-out roll. One might be tempted to conclude, therefore, that the odds of winning the come-out roll with a don't-pass bet are 50/50. One would be mistaken. Probability is more complicated than that, even for a relatively simple game like craps, which is why the field has fascinated scientists and mathematicians for centuries.

CHANCE ENCOUNTERS

Among the first to analyze games of chance with an eye toward odds and winning strategies was a sixteenth-century physician, astrologer, and mathematician named Gerolamo Cardano. Born in 1501, his was not the most auspicious of beginnings. His mother, having already borne three children and clearly being fed up with parenthood, tried to abort him with a brew of wormwood, barleycorn, and tamarisk. Gerolamo survived but promptly contracted bubonic plague when he was just a few months old—usually a death sentence at the time, particularly for an infant. Astoundingly, he survived that, too. (His wet nurse and three half-brothers perished.)

His father, Fazio, wanted the teenage Gerolamo to study law, but the boy longed to study medicine instead. He initially supported his studies by tutoring others in geometry, alchemy, and astronomy, as well as casting horoscopes. (Astrology and alchemy were still considered legitimate fields of study.) But then he developed a taste for gambling and found he had a gift for beating the odds. He quickly amassed winnings of 1,000 crowns, more than enough to pay for his education, and in 1520 began writing a treatise, *The Book on Games of Chance*, which he kept revising right up until his death.

Cardano was a better gambler than a physician, it seems—or rather, he lacked the business acumen to market himself to prospective patients. He struggled mightily to support his family early in his career, and soon found himself resorting to gambling again to make ends meet. Eventually Fortune seemed to smile on him: He published a series of successful books and by

1550 became the renowned physician he'd always dreamed of being.

If only he hadn't had children. Cardano's appalling offspring were a trio of bad seeds whose behavior would make Caligula blush. His daughter Chiara seduced her older half-brother, Giovanni, at the age of sixteen, became pregnant, aborted the fetus, and continued to philander even after her marriage, eventually contracting syphilis. That same brother was later convicted of poisoning his wife; Cardano spent a fortune on his defense, to no avail. Giovanni was summarily executed, most likely deservedly so. The younger son, Aldo, became a torturer for the Spanish Inquisition, testifying against his own father so that Cardano briefly landed in jail. Cardano finally died in September 1576, penniless and quite mad, having burned more than half of his manuscripts before shuffling off this mortal coil.

Among his surviving manuscripts was *The Book on Games of Chance*, finally published in 1663, almost a century after Cardano's death. By that time, others had replicated and outpaced Cardano's analysis, but the beleaguered physician with the rotten luck deserves his minor place in the annals of probability theory. In chapter 14, titled "On Combined Points," Cardano laid out what we now know as the law of the sample space. The sample space is simply the set of all possible outcomes of a random process (like the roll of the dice or flipping a coin). Cardano reasoned that the probability of winning a roll of the dice, for example, is equal to the proportion of winning outcomes. A die can land on any one of its six sides, and those six potential outcomes make up the sample space. Place a bet on one such number, and your chance of winning is 1 in 6; place bets on three such numbers, and your odds improve to 3 in 6.

His methodology served him well as a gambler, but Cardano's analysis was rather flawed. He assumed that all outcomes were equally likely; in fact, different outcomes have different probabilities. Galileo Galilei demonstrated this in the early seventeenth century in a short paper entitled "Thoughts About Dice Games." Galileo wasn't especially interested in probability theory, preferring to roll balls down inclined planes and time their rate of travel. But his patron, the Duke of Tuscany, was an inveterate gambler and thus keenly interested in the question of why—for games played with three dice—the number 10 seemed to occur a tad more frequently than the number 9. Galileo concluded (correctly) that this occurred because there were more combinations that yielded a 10 than yielded a 9. There are twenty-seven ways to roll a 10 with three dice, compared to twenty-five possible combinations for a 9. It's now an established tenet of probability theory that the odds of a particular outcome are dependent on the number of ways in which it can occur.

Galileo took his analysis no further; his research interests lay elsewhere. Yet wealthy and titled patrons with gambling problems continued to push for advances in probability theory, most notably a social-climbing French essayist named Antoine Gombaud, who adopted the title chevalier after the character in his many dialogues who represented the author: Chevalier de Mere.

Gombaud was a man of letters who fancied himself an amateur mathematician, and in 1654 he found himself pondering what is known as the problem of points: How do you determine how the stakes in a game of chance should be divided if, for some reason, the players were interrupted and never finished their game? It was first proposed in 1494 by an

Italian monk named Luca Pacioli in his treatise *Summa de arithmetica, geometria, proportioni et proportionalita.* (Yes, even monks fell victim to the lure of gambling. They didn't have television in the Middle Ages.) So this question had been knocking around gambling circles for nearly two hundred years by the time Gombaud decided enough was enough—he wanted a solution to the conundrum.

Gombaud turned to a young mathematician named Blaise Pascal, who had taken up gambling when his doctors advised him to abandon mental exertions for the sake of his health. Pascal suffered from chronic stomach pain, nausea, migraines, and partial paralysis of the legs, among other ailments. Intrigued, Pascal quickly realized he would need to invent an entirely new method of analysis to solve the puzzle, because the solution would need to reflect each player's chances of victory given the score at the time the game was interrupted. Thus began his legendary correspondence with fellow mathematician Pierre de Fermat, which over the course of several weeks, laid the foundation for modern probability theory. They quickly realized that in order to solve the problem it would be necessary to list all the possibilities and then determine the proportion of times that each player would win.

Caltech mathematician Leonard Mlodinow gives one of the clearest explanations of how to solve the problem of points in his book *The Drunkard's Walk*, using the example of the 1996 World Series, in which the Atlanta Braves beat the New York Yankees. Atlanta won the first two games, but what were the odds of a Yankee comeback at that point? To get the answer, you would need to count every scenario in which the Yankees could have won and compare that to the number of scenarios in which they could have lost. By that reckoning—which assumes that the Yankees and the Braves had equal chances of winning each subsequent game—the chance of an overall Yankee victory would have been 6 in 32, or around 19 percent, compared to 26 in 32, or about 81 percent, for an Atlanta victory. "According to Pascal and Fermat, if the series had been abruptly terminated, that's how they should have split the bonus pot, and those are the odds that should have been set if a bet was to be made after the first two games," Mlodinow concludes.

Pascal's bank account may have suffered during this period of his life, but his health was never better. Ironically, the mental exertions of his correspondence with Fermat triggered a "trance" a few weeks later, from which Pascal never fully recovered. He became deeply religious, eschewing his former "corrupt" ways, and died of a brain hemorrhage at thirty-six. Maybe he should have stuck with gambling.

RISK AND REWARD

What happens in Vegas stays in Vegas, or so the advertising tagline goes—and more often than not, your money stays too. For craps, the probabilities are fairly straightforward because there are only two dice with six sides each, so there are only 36 possible combinations: six possibilities for each of the two dice ($6 \times 6 = 36$). Yet not all outcomes are created equal, and therein lies the secret of the house advantage. There are more ways to roll a 7 than a 2, for example. To roll a 2, you would need to roll snake eyes ($1 + 1$). In contrast, there are three different combinations that total 7: $1 + 6$, $2 + 5$, and $3 + 4$. Furthermore, because each die is distinct, probability also includes the combinations 4 + 3, 5 + 2, and 6 + 1. Ergo, 7 is the most likely number to be rolled. It's no accident that the losing roll (craps) is 7 once the game gets under way.

How does this play out when one takes into account the odds and payoffs for the various bets in craps? For the pass and don't-pass bets, the payoff odds are one to one: Winners receive one dollar for each dollar they bet. This does not mean you have a 50/50 chance (0.5) of winning a pass-line bet; the actual probability is exactly 0.49293—just slightly less than a 50/50 chance, giving the house an edge of about 1.42 percent.

Once the point has been established, the next most favorable side bet to further narrow the house's edge is called a "free-odds" bet. For example, before the next roll, S-Money would add between one and three extra chips behind his original pass bet. While line bets have a one-to-one payoff, the payoff for a free-odds bet is determined by the exact mathematical odds against winning the bet. If the point is either a 4 or a 10, the odds against winning are 2 to 1; ergo, the payoff if the shooter rolls the point before a 7 is 2 to 1. So if S-Money bet $10 as a free-odds bet, he would win $20. If the established point is a 5 or a 9, the odds against winning are 3 to 2, so the same $10 free-odds bet would win $15. And if the point is a 6 or an 8, the odds against winning are 6 to 5, so a $10 free-odds bet would bring in $12.

Dominic assures us that the free-odds bet gives the house the smallest possible advantage, and thus is an excellent way to maximize one's winnings. But there's still just a single number (the point) by which you can win. You could increase your chances of winning if there were more possible winning numbers. That's where the "come" bets and "point" bets come in. For a come bet, you place a chip in the come section of the table before the next roll, and whatever number the shooter rolls becomes a new point for that particular chip. So now you can win on this point or on the original point established on the come-out roll. If the shooter rolls a 7, of course, you lose outright—because the game is over.

There are some disadvantages to the come bet. First, it is a "contract bet," meaning it remains in place until the end of the game, just like the pass and don't-pass bets. (You can also place don't-come bets, once again betting against the shooter.) Second, you are at the mercy of a roll of the dice to determine the new point. If you want to be able to pick your own number to be the new point, and have the freedom to add or remove chips at will, you should make a point bet. The payoff odds aren't quite as good as the come bets, but you have more control over the board. There are plenty of other, higher-risk bets, but these are really the only bets you can make in craps with reasonable odds. You'll still lose money in the long run, but you'll lose it much more slowly.

Now it's time to play the game for real. The casino graciously sets up a small-stakes table just for the newbies for one hour so we can practice our newfound skills. S-Money being the more math-savvy in our household, he shrewdly opts to take Dominic's advice and maximize the size of his odds bets in relation to his line bet, thereby reducing the house edge to a whisker (although not eliminating it entirely). Because we're experimenting, I choose to focus on placing come bets and point bets, trading better payoff odds for more control over the board, to see how these two strategies compare.

From a psychological standpoint, craps is an ingeniously designed game. The odds are certainly rigged in the casino's favor, but they are not rigged too heavily in that direction. That would be no fun at all. Players need a sense of reward, even if it's just the illusion of winning once in a while. That's what makes the game so addictive. I soon notice that we do win rolls, sometimes several in a row, but in all but a few cases, the money we win never quite adds up to the money we spend placing even the minimum bets. The result: At best, playing craps is a slow bleed that one can easily not notice, because the tiny amounts won in between losses blind players to the long-term financial hemorrhage that is taking place.

Yet somehow we walk away one hour later with combined winnings of $145. (I win $45, S-Money wins $100—a neat illustration of how the various odds play out, albeit purely anecdotal.) My interpretation is that we hit a streak of good luck, and had the sense to quit playing while we were still ahead. Ever the physicist, S-Money loftily informs me that, according to probability theory, "hot" or "cold" streaks are merely a perception. Each roll is independent of the previous and subsequent rolls; that is the nature of true randomness. So the odds are the same for each roll, even if the shooter has rolled the point twenty or two hundred times in a row; the outcome of the last roll does not affect what happens next. There is no such thing as being "due" for a win (or a loss).

Still, it is possible to figure out how often we are likely to have a winning session of craps. Translating this concept into actual calculus is tricky, in part because throwing dice falls into the realm of discrete events—analyzing event probabilities under repeated trials—whereas calculus, by definition, deals with continuous things. If you plot the probability of the outcomes of individual rolls, you'll get a shape resembling a pyramid—a perfectly good shape, but not one that represents a continuous function.

However, if you throw the dice two thousand times (or

more), add up how much you win and how much you lose each time, and plot it all out on a Cartesian grid, the result is a standard bell curve, also known as a normal distribution curve. For any random sample—say, many random rolls of the dice—you will get a distribution of values clustered around an average (or "mean") value. That mean value is the peak, or highest point, of the curve, where the most data points cluster together; there are fewer and fewer data points as we move out to the edges. In craps, big wins or big losses happen very infrequently, and would be found at the extreme edges of the bell curve, while outcomes with smaller wins and losses would cluster near the peak of the curve.

Now that we have a pretty bell curve, we can use calculus to determine how often we will have a winning session in craps. First we need to understand what it is we are calculating; we have to set up the story. Every time we throw the dice, the

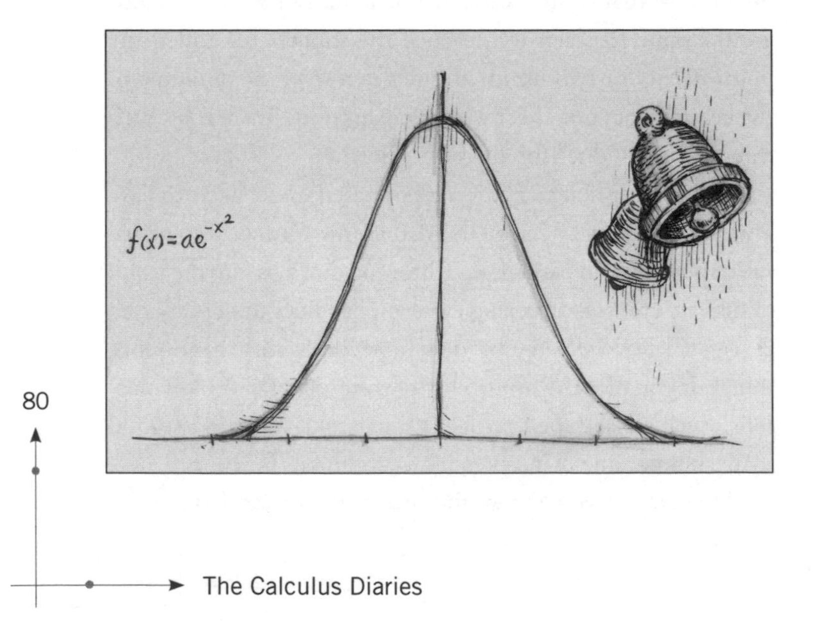

probabilities for rolling a specific number don't change from the odds outlined above; there is still a 1 in 6 chance of rolling a 7 with each and every roll, as represented by the pyramid. We are asking a different question: Given that we know we will throw the dice two thousand times, what is the probability that we will win or lose?

This reduces the question to an either-or option, assuming even odds—and remember that the game of craps is *not* even odds; we're just making that assumption for the sake of simplicity. In this case, for every roll, there is a 50/50 chance that we will win or lose, and each outcome is separate from those before and after. It's known as a random walk or, as many mathematicians like to call it, the drunkard's walk. The probability that any one session of craps is an overall win or a loss approaches the distribution of this smooth bell curve the more times we throw the dice. If we throw the dice an infinite number of times, our win-loss rate will match the bell curve exactly.

How do we determine that likelihood? We take an integral. The actual formula for the integral from one point to another on a bell curve has never been explicitly written down; it is usually calculated with a computer. But here is the gist of the concept. Imagine a number line that runs from negative infinity (on the left) to infinity (on the right), with 0 smack in the middle, and a standard bell curve peaking at 0. This represents the distribution of outcomes for a 50/50 chance of winning or losing. The probability of losing will be the area under the curve that spans from minus infinity to 0, while the probability of winning will be the area under the curve from 0 to infinity. Each is equal to one half in this simplified example. The more times we roll the dice, the closer we will come to matching those probabilities. With 50/50 odds, for an infinite number of rolls, we will break even.

We can be even more specific by picking a random point on the *x* axis—say, 500—to determine the likelihood that we will either lose money or win up to $500. The answer will be the area under that portion of the curve that runs from negative infinity to 500. If we want to know the likelihood that we will win more than $500, we determine the area under that portion of the curve that runs from 500 to infinity.

The biggest problem when it comes to craps is that the odds are not 50/50. Let's say the house has a slight edge, making the odds 49/51. Now our bell curve is shifted slightly to the left on our grid, making it slightly more likely that we will lose; and the longer we play, the closer we will get to that distribution. We also need to specify the size and type of bet for each roll, because the probabilities in craps are linked not just to the outcomes of the rolls of the dice, but also to the payoff rates for different kinds of bets.

GAMING THE SYSTEM

We won at craps because we got lucky in the short term: We hit a probabilistic sweet spot by pure random chance and had the sense to quit while we were ahead. Vegas notoriously attracts gamblers convinced they have discovered a "system"—a perfect strategy to beat the house. They are deluding themselves. Even assuming these perennial optimists have taken every single variable into account for their calculations, it takes only the tiniest house advantage to tip the scales irrevocably. We played for just one hour. Play the game long enough, and eventually you will lose everything. The occasional perceived hot streak or lucky break doesn't alter that fact. The casinos are very up front about this. Another craps dealer in the New York, New York casino—let's call him Vito—didn't mince any words on that score: "Everyone thinks they got a system. You think you're gonna beat this table? Go ahead and try. We got ATMs all over the casino, just for people like you." Listen to the wisdom of Vito, my friends. Forewarned is forearmed.

Even if the odds are in your favor, there's no guarantee you'll win. Let's imagine the situation were reversed, and the *players* had the slightest advantage; it wouldn't necessarily translate into an automatic win. You must pay just as much attention to your bankroll as the odds of winning; if the odds are good but you're betting a substantial portion of your bankroll on each roll of the dice, it's enough to wipe out any advantage pretty quickly. That's the essence of a little exercise called gambler's ruin. It's a favorite of University of Washington physicist Dave Bacon—better known to the blogosphere as the Quantum Pontiff—who became fascinated with cataloging the outcomes of repeated throws of dice as a child. He admits this made him an übergeekazoid, but it probably saved him a lot of money in the long run.

Gambler's ruin begins with the assumption—a false one, when it comes to craps—that the player has a slight advantage in a game of chance and should win slightly more than half the time. Say you have a bankroll of x dollars. For every dollar you bet, you win another dollar or lose the original dollar, depending on the outcome. (This is the same payoff rate as the pass and don't-pass bets in craps.) What is the probability that you will run out of money, even with that slight advantage, rather than increase your bankroll by, say, doubling your money?

Drawing on that childhood fascination, Bacon devised a handy formula and plugged in a few values to see if a pattern emerged. He found that even with a fairly large advantage—say, 55/45—if you only start with $10 and make a fixed bet each time, there is an 11.8 percent chance of being ruined before you succeed in doubling your money. If you have a 51/49 advantage and a starting bankroll of $178, your chances of ruin before doubling up decrease to 0.1 percent, or 1 in 1,000. In craps, of course, you don't have an advantage. Bacon has crunched those numbers, too. If the house has the usual edge of 1.42 percent, and you start with $100 and want to double it, your probability of ruin is 98.2 percent. That's why casinos make such a killing.

So the first rule of gambling, for those who have studied the odds, is simply, Don't.* Still, craps is quite a lot of fun, provided you view it as harmless entertainment, rather than a get-rich-quick scheme to pad your 401(k). A good rule of thumb is to budget a set amount you are willing to lose and just chalk it up to the price of a day's entertainment. Once you lose that amount, suck it up and walk away, and maybe explore a few of the other delights of Vegas.

Admittedly, this is easier said than done. For one thing, casinos employ a stickman at every craps table, whose job is to talk up the game and encourage players to make the riskier

---

* Legend has it that the American Physical Society once held its annual meeting in Las Vegas. The assembled physicists shunned all the usual decadent delights: showgirls, hookers, blackjack, roulette, craps, and copious amounts of alcohol, plus they were lousy tippers. There wasn't a single barroom brawl. The city made so little money, the APS was asked never to come back to Vegas. Now the society holds its major meetings in more sober, straitlaced places like Cincinnati, Indianapolis, and Denver.

bets. For another, a 2008 paper in the *Journal of Marketing Research* reported on a study by two professors at the University of California. They found that even if people went into the casino determined to stay within their gambling budget, the pain of losing would usually cause them to bet more money in hopes of recouping their losses. Those who won tended to keep to their budget.

If someone develops a gambling addiction, the problem is even worse. In 2007, a Nebraska businessman named Terrance Watanabe lost nearly $127 million in a yearlong binge at the Caesars Palace and Rio casinos, blowing most of his personal fortune. When parent company Harrah's Entertainment sued him for nonpayment of his gambling debt, Watanabe countersued, claiming casino staff plied him with drinks and encouraged him to gamble while intoxicated, thereby impairing his judgment. There could be an element of truth to that: High rollers like Watanabe—"whales" in the jargon of casino staff—are a lucrative source of income for casinos. As such, casinos treat them very well, doling out all manner of luxuries, free of charge, to keep them happy. But there are rules: Nevada gaming regulations stipulate that someone who is clearly intoxicated should not be allowed to gamble. In fact, Watanabe claimed he was barred from the Wynn casino for compulsive drinking and gambling; the Harrah's establishments welcomed him with open arms.

Watanabe's spectacular downfall is a rare occurrence. It is certainly possible to maximize your fun playing craps in a casino without breaking the bank. Just follow this basic principle: You want to play as long as possible with a fixed amount of money. That means losing as little as possible with each bet by choosing those bets with the most favorable odds and payouts. It's easy to determine the optimal percentage of your bankroll you should bet in order to maximize your long-term return without busting out, using something called the Kelly criterion.

Born in Texas, John L. Kelly was a Naval Air Force pilot during World War II who survived a plane crash into the ocean and eventually earned a PhD in physics from the University of Texas–Austin. He found work in the oil industry, using his scientific training to identify likely oil sites. But his employer's instincts were better than Kelly's models, so Kelly decided the oil business was best left to those with a nose for hidden deposits and found himself working for Bell Labs, one of the most prestigious research centers in the United States. He cut a colorful figure among his fellow physicists, with his Texas drawl, passion for guns, and penchant for taking calculated risks.

It was a hugely popular television game show called *The $64,000 Question* that inspired Kelly to devise his famous formula in the 1950s. People would place bets on the most likely contestants to win. But there is a three-hour time difference between New York City—where the show was produced and aired live—and the West Coast. Kelly heard a rumor that one gambler on the West Coast had a partner back east tell him the winners by phone so that he could place bets before the show aired in the West, giving said gambler an inside track. This spurred Kelly to ponder probabilities and gambling. He reasoned that if a gambler with an inside track bets everything he or she has on the basis of those tips, the gambler will lose everything the first time he or she gets a bad tip. But if the same gambler makes just the minimum bet for each tip, that insider information no longer confers much of an advantage. Recognizing the importance of how much someone bets in fashioning a winning strategy, Kelly determined that dividing your edge by the odds tells you what percentage of your bank roll you should bet each time.

The odds determine how much profit you make if you win; the edge describes the amount you expect to win on average if you make the same wager repeatedly under the same probabilities. Remember the lesson of gambler's ruin: Even if the odds are in your favor, you still don't want to bet your entire bankroll in one fell swoop; your odds of losing everything on one roll are much higher. Play it safe and bet too little, however, and your return won't be sufficient to make up for the inevitable losses. Kelly's formula reveals the optimal betting strategy for maximizing long-term returns. For a bet with even odds, Kelly tells us to bet a fraction of our bankroll that is determined by $2p-1$, where p is our probability of winning.

When it comes to playing craps in a Vegas casino, it will be a discouraging answer unless you have the good fortune to be the house. Players usually have an edge of zero at best (a 50/50 chance) and more often it is slightly less. In either case, the Kelly criterion says that the best way to maximize your long-term return in craps is to bet 0 percent of your bankroll—that is, not to play. But that is just a detached, mathematical analysis that doesn't take into account the fun factor, the sheer pleasure one derives from playing craps.

We can tweak this problem a little to take that subjective quality into account by assigning it a quantitative value: Let's say the odds are 49/51, giving the house a 2 percent edge, but the fun factor is 3 percent, giving us a net edge over the casino of 1 percent. That corresponds to a winning probability of 0.51, so the Kelly criterion tells us to bet 2 percent of our bankroll. Now, we can place our bets accordingly to optimize our fun—that is, play as long as possible by maximizing our long-term gains. We'll still most likely lose in the end, but we will be getting the most bang for our buck.

There is a downside to the Kelly criterion, or rather, a kind of trade-off: Following the Kelly criterion exactly leads to a lot of volatility in the outcomes. In the long term, it works; in the short term, it can lead to intense anxiety over the wild fluctuations in one's fortunes. For those who prefer a bit less drama in their gambling, a popular middle-ground strategy is to bet half of what the Kelly criterion recommends. This optimizes your return to within three fourths that of the Kelly criterion while greatly reducing the volatility. It seems apt that the optimal formula for long-term gain would increase short-term risk, considering that the man himself was a bit of a daredevil. Ironically, Kelly never actually put his method to the test: He died of a brain hemorrhage in 1965 at the age of forty-one while walking down a Manhattan sidewalk. What were the odds of that?

NEEDLE IN A HAYSTACK

We round out our wild Vegas weekend with several hours of good old-fashioned poker—a game of skill and strategy, as opposed to pure random chance, wherein the casino takes a cut of the pot instead of relying on a built-in house edge. What have I learned? "Craps" is an apt moniker. Also? I can't bluff worth a damn at Texas hold 'em. But in the end, we emerge from the weekend with our wallets relatively unscathed.

Relaxing over cocktails at the Bellagio that evening, I ponder the fact that probability theory and gambling are also linked to fortune-telling and one of the most famous "natural" numbers, π. In Philip Pullman's *The Amber Spyglass*, fictional Oxford physicist Mary Malone finds she can communicate with the mysterious, conscious particles collectively named Dust using the yarrow-stick casting methods of the I Ching (it's also possible to use coins). For those who scoff that a physicist would never express any appreciation for a "supernatural" method of divination, consider this: When he was knighted, Neils Bohr included the yin-yang symbol in the design for his coat of arms, to reflect his appreciation for the I Ching's ingenious use of probabilistic concepts.

Mary Malone's divination method has a real-world counterpart in one of the oldest problems in geometrical probability, known as Buffon's needle. This experiment was the brainchild of a French naturalist and mathematician named Georges-Louis Leclerc, Comte de Buffon. Born and raised on the Côte d'Or, the young George-Louis started off studying law before getting sidetracked by mathematics and science. It's not clear that he ever earned a degree, because he was forced to leave the university after getting tangled up in a duel. He then toured Europe, only returning when he heard his father had remarried—not so much out of familial devotion as concern over collecting his inheritance.

Buffon fils is best known for writing the *Histoire naturelle*, a whopping forty-four volumes of encyclopedic knowledge that covered everything known at that time about the natural world. A full hundred years before Charles Darwin's *Origin of Species*, Buffon noted the similarities between humans and apes and mused on the possibility of a common ancestry, concluding that species must have evolved since that common point. He never proposed an actual mechanism for this evolution, but his tome was translated into numerous languages and certainly influenced Darwin, who described Buffon—in the foreword to the sixth edition of *Origin*—as "the first author who in modern times has treated it in a scientific spirit."

Buffon's quirky contribution to probability theory lies in a paper he published in 1777 entitled, *Sur le jeu de franc-carreau* [On the Game of Open-Tile]. He first considered a small coin—an ecu, for all you crossword puzzle buffs—thrown randomly on a square-tiled floor. It was all the rage in Buffon's social circles to place bets on whether the coin would land entirely within the bounds of a single tile or across the boundaries of two tiles right next to each other. Buffon had a bit of an advantage over his peers thanks to his mathematical interests. He realized he could figure out the odds of the wager using calculus, making him the first person to introduce calculus into probability theory.

Buffon noted that the coin would land entirely within a tile whenever the exact center of the coin landed within a smaller square—and that smaller square's side was equal to the side of a floor tile minus the diameter of the coin used in the toss. He concluded that the probability of the coin landing entirely inside a single tile could be expressed mathematically as the ratio of the area of the tile to the area of the smaller square.

Buffon performed the same experiment using a sewing needle and a checkerboard—hence the name Buffon's needle. Drop the needle onto the checkerboard, and one of two things happens: Either the needle crosses or touches one of the lines, or it doesn't cross any lines. (This assumes parallel lines or squares spaced about one inch apart, and the use of a needle one inch long.)

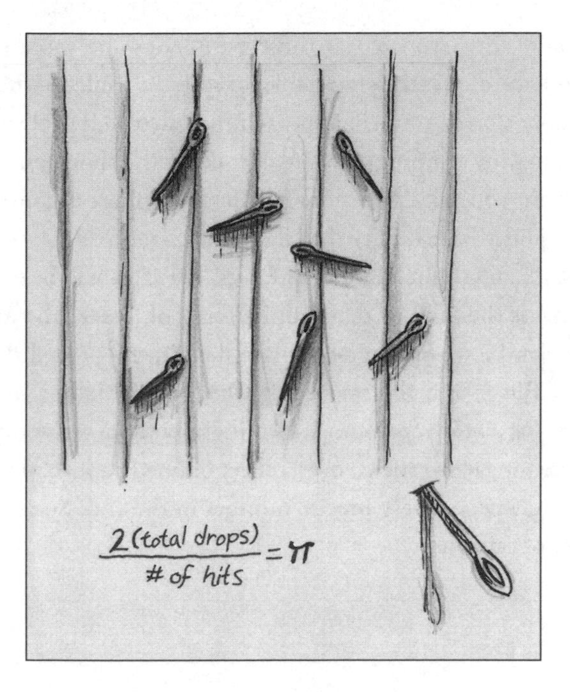

Buffon dropped the needle over and over again, keeping track of how the needle randomly landed each time. His found that the probability that a dropped needle (or tossed coin) would cross a line is approximately 2 divided by π. He divided the number of crossing needles by the total number of needles, and realized that the more times one drops the needle, the closer one would approach the value of the probability—that is, the closer one would come to the value of π.

There are many online versions of this experiment, wherein the player can repeat the "toss" as many times as desired: five hundred, a thousand, even a hundred thousand times. Once again, the more times you repeat the experiment—the more times you roll the dice at the craps table, or spin the roulette wheel—the more closely you will approach the calculated probability. There may be winning or losing streaks in the short term, but the more you play, the more predictable things become. It's just a quirky little oddity that the value relates to π.

With an infinite number of tosses, the value will be exactly π—that is the limit of that infinite series of tosses. The mathematician Pierre-Simon de Laplace definitively proved this in 1812. This is also the essence of what Mary Malone discovers in *The Amber Spyglass*. A seemingly random scattering of needles (or yarrow sticks) over a sheet of lined paper can nonetheless give you a very precise number in the end. Such is the power of calculus.

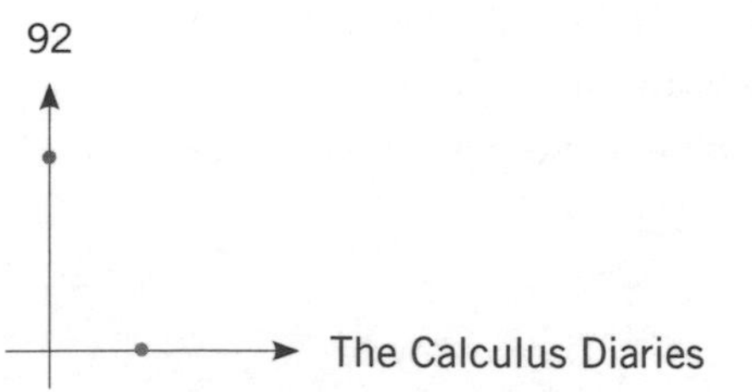

The Devil's Playground

> Mechanics is the paradise of mathematical science because here we come to the fruits of mathematics.
>
> —LEONARDO DA VINCI

It is a bright and sunny Sunday afternoon inside Disneyland's California Adventure theme park. Visitors meander blithely through the broad "streets," nibbling on ice cream and occasionally pausing for photo ops with life-size characters from popular animated features like *Monsters, Inc.*, *The Incredibles*, or *Lilo & Stitch*. They seem oblivious to the ominous shadow cast by the Tower of Terror, looming nearly two hundred feet above the ground, or the screams emanating from within the structure. Blackened scorch marks decorate the crumbling facade, where lightning supposedly struck in 1939, with tragic results.

Of course, nothing in Disney's many theme parks is real. Those are screams of exhilarated delight, not abject terror, piercing the grim walls. Inspired by the classic TV series *Twilight Zone*, the Tower of Terror is Disney's theatrical twist on the classic free-fall ride. We've been waiting in line for nearly forty-five minutes to experience those few fleeting moments of thrills and chills.

Inside we encounter the faded glory of a bygone era: Sagging overstuffed furniture, layers of dust, cracked plaster, and glass chandeliers laced with fake cobwebs grace the "lobby." We gradually shuffle our way to a boarding dock for mock elevators, where an employee dressed as a bellhop ensures we are all tightly strapped into our seats. Our elevator rises midway to the top and stops, and we are treated to Rod Serling's disembodied voice regaling us with the saga of a dark and stormy night on October 31, 1939, when five hotel guests stepped into an elevator and were launched into . . . the Twilight Zone!

Before we can snicker at the cheesy effects, our elevator makes a sudden gut-churning drop back to the ground floor and then shoots up all the way to the top of the structure (ostensibly the thirteenth floor), the acceleration pushing us into our seats. We pause just long enough to glimpse the rest of the park spread out two hundred feet below, before the elevator car plunges again—one short drop followed by one longer drop, each accompanied by a glorious moment of weightlessness. Then we hurtle back up to the top for one final free fall back down to the "basement," where the faux bellhop waits to usher us back out into the Southern California sunshine. As the ride ends, Sean turns to me and gleefully exclaims, "Hey! We made a parabola!"

Everyone should visit Disneyland with a physicist in tow, just for the novelty; it's an entirely new way of looking at the Magic Kingdom. (Motto: "All headgear is improved by the addition of mouse ears.") I guarantee that nobody else on that ride found their thoughts wandering to calculus and parabolic curves; they were too busy screaming with joy at the free fall. Sean had never been to Disneyland, and I decided it was time to rectify that gap in his cultural development, insisting that it is a slice of Americana that must be experienced firsthand—and besides, what better place to find examples of calculus and classical mechanics in action?

Amusement-park physics is all the rage among high school physics teachers desperate for novel ways to engage their easily distracted young charges. Case in point: Every year, on Physics Day, more than four thousand high school students swarm Six Flags America in Largo, Virginia, armed with homemade accelerometers (devices to measure acceleration) and stopwatches, eager to experience the park's extreme roller coasters—and perhaps learn a little physics along with the adrenalin rush. So Sean good-naturedly agreed to spend a Sunday at Disneyland, being dragged from one long queue to another, filled with overexcited youngsters, frazzled parents, and purple-haired hipsters with multiple piercings doing their damnedest to look bored and act as though they were Really Just There for the Irony.

FREE-FALLIN'

That diverse mix of young and old is exactly what Walt Disney had in mind when he first dreamed up the notion of a "magical park" in the late 1930s. World War II put his plans on hold, but by 1953, he had found one hundred acres just outside Los Angeles where he could build his Magic Kingdom. On July 19, 1955, Disneyland held its official grand opening. It was a disaster. Disney had intended the day to be an exclusive, invitation-only event for a select 6,000 people. But counterfeit invitations were quickly forged and snapped up by eager hordes. People began lining up at the park gates as early as two A.M., and by midafternoon, over 28,000 "ticket holders" had swarmed the park. Vendors ran out of food, and all the rides were overcrowded. A few desperate parents tossed their wailing offspring over the shoulders of bystanders blocking the way, just to get them onto the King Arthur Carousel.

The weather didn't cooperate either. The mercury hit 110 degrees Fahrenheit, part of a fifteen-day heat wave that baked the greater Los Angeles area that July. Newly laid asphalt hadn't had time to set, so women's high heels got stuck in the melting tar, and hardly any of the park's water fountains worked because of an ongoing plumber's strike.* Adding insult to injury, there was a gas leak that forced the afternoon closure of Adventureland, Frontierland, and Fantasyland; only Tomorrowland emerged from the debacle unscathed. But Disneyland proved hugely successful in the long run. By the time the park celebrated its tenth anniversary in 1965, over 50 million people had visited.

Disneyland is much better at crowd management these days, even though lines remain long for the most popular rides. And the Disney empire has expanded and gone global. Within the original park, in Anaheim, there is now New Orleans Square,

* Apparently Disney was forced to choose between working water fountains or running toilets, and he wisely chose the latter. That didn't stop the ungrateful crowds from accusing him of deliberately sabotaging the water fountains to sell more soda (Pepsi had sponsored the park opening).

Critter Country, and Mickey's Toontown, in addition to the original four "lands." Florida has Disney World, and there are now Disney theme parks in Paris, Tokyo, and Hong Kong. The California Adventure theme park opened adjacent to Disneyland in 2001; the Tower of Terror can be found in the Hollywood Pictures Backlot section of the park.

Human beings have thrilled to the sensation of free fall for centuries, with occasionally dire results. Witness the enormous popularity of bungee jumping, which has its roots in the ancient Aztec ritual of the Danza de los Voladores de Papantla; the *danza* is still practiced today by "Papantla flyers." In the 1950s, British documentary filmmaker David Attenborough took his BBC film crew to Pentecost Island in Vanuatu, where they recorded several young tribal men who jumped from tall wooden platforms with vines tied to their ankles as a test of courage. It was only a matter of time before extreme sports enthusiasts had the brilliant notion of harnessing themselves to bungee cords and jumping off tall structures for fun (and the occasional profit).* Bungee jumping quickly spread around the globe, despite numerous accidents and the odd fatality.

For those (like me) who prefer a more sedate form of thrill-seeking, there are mechanical free-fall rides with, shall we say, more rigorous safety constraints. Six Flags Great Adventure introduced one of the first true free-fall experiences in 1983. The L-shaped structure featured a four-passenger car lifted via

---

* A group of British adrenalin junkies formed the Oxford University Dangerous Sports Club and leaped from Bristol's 250-foot Clifton Suspension Bridge in 1979. They were promptly arrested, but undeterred: They went on to jump from the Golden Gate Bridge, mobile cranes, and hot-air balloons.

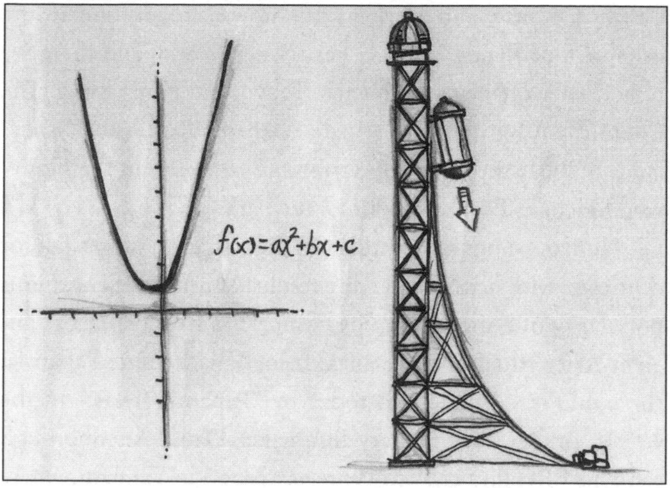

hydraulics to the top of a 130-foot tower and suspended for a few seconds. At the buzzer, the car would plunge down the drop track and onto the horizontal exit track to end the ride. The latter was necessary because coming to a sudden stop at the end of the drop would most likely cause serious injuries. The deceleration period dissipates all that kinetic energy over a longer period of time so it isn't transferred all at once to the passengers.

The Tower of Terror is a variation of a "drop tower" ride that gradually has replaced the classic free-fall design since the 1990s, largely because it is closer to a true free-fall experience, and there is less mechanical wear and tear. A gondola or car—in this case, the mock elevator—is propelled upward toward the top of a large vertical structure and then falls back toward the ground. The brakes kick in before impact, slowing the ride, although the

Tower of Terror essentially "bounces" its riders a few times before finally coming to rest.

Technically, we enter free fall when there is no longer any force (other than gravity) acting directly on us. Think of tossing an apple into the air. The moment it leaves your hand and you stop applying that upward force, it is in free fall. It continues traveling up, moving more slowly as gravity overpowers its upward motion, has a brief moment of hang time (that period of weightlessness), then begins its descent. Our car in the Tower of Terror follows the same trajectory. It receives an initial push from the hydraulics, but at some point that force is removed and we finish our ascent using pure momentum. That brief, exhilarating period of weightlessness occurs because riders fall at the same rate as their surroundings—in this case, their seats. NASA's infamous "vomit comet" follows a parabolic trajectory while in flight, taking such extreme lifts and dips that it can achieve about twenty to thirty seconds of weightlessness for every sixty-five seconds of flight.

Thrills aside, the Tower of Terror provides an excellent example of calculus as it applies to classical mechanics. True, our physical motion is straight up and down. But if we plot our change in height (position) over *time* point by point on a Cartesian grid—both ascending and descending trajectories—and connect the dots, we end up with the telltale parabolic curve that so delighted Sean at the ride's end. (The same is true for the apple.)

How does this work? We begin with our starting velocity. It is not 0, because we are specifying our starting velocity at the moment we enter freefall, not at the start of the ride (when it would be 0). We need a speedometer to tell us how fast we are traveling at that moment, and that number becomes our starting velocity (a constant). Let's imagine the Tower of Terror tracks acceleration for us. We can take an integral of our acceleration to get our velocity, essentially adding the acceleration—in this case, the gravitational constant—at each moment in time.

The result, when we graph it out, is a straight downward-sloping line. This is our velocity function. We can use that to determine our position (height) by taking *another* integral, adding together how far we traveled at each point in time. Plot each position as a function of a time and you get a pretty parabola. Now that we have a position function, finding our height at any given point in time is a snap.*

Many years ago, I went to Six Flags New Jersey with a group of friends, and we all went on the Devil Dive—a cross between bungee jumping and a really big tire swing. Three of us were strapped into one big harness and lifted to the top of a 200-foot tower. That might not sound very high unless you happen to be one of the people hanging precariously at the top of it; then terra firma seems a very long way down. One of my cronies had just enough time to nervously remark, "Um, maybe this wasn't such a good idea after a—AUUGHH!" The buzzer sounded, the catch released, and we plummeted, screaming, toward the ground.

Just as we were about to hit the ground, the harness caught and swung us outward in a sweeping pendulum motion, moving through space along the trajectory of an arc of a circle. We swung back and forth like a three-person pendulum, until we

* Check out appendix 1 for the mathematical solution to this problem.

slowed down sufficiently for the ride operators to grab us and release us from the harness.

The Devil Dive gives us a double dose of Galileo. First, there is the free fall. It is roughly the same problem outlined above, except in this case our position as a function of time forms only half a parabola, because we don't enter true free fall until we begin our descent. Our acceleration is –32 feet per second per second at any time (t) after our drop begins. (The sign is negative because we are falling and our height is decreasing.) We can take an integral to get our velocity, and then integrate the velocity to get our position function, just as we did before.

Second, there is the pendulum motion at the end of the ride. An oft-told anecdote from Galileo's youth tells of the seventeen-year-old future scientist growing bored during Mass in a drafty cathedral in Pisa. He noted a chandelier hanging from the ceiling swaying in the breeze. Sometimes it barely moved; other times, it swung in a wide arc. This proved more interesting to the teen than the priest's sermon, and he began timing the swings with his pulse, with a surprising result: It took the same number of beats for the chandelier to complete one swing, no matter how wide or narrow the arc. Granted, the chandelier moved faster during wider arcs, but it completed its arc of motion in the same amount of time. The same motion can be seen in playground swings and the arc we make at the end of our Devil Dive. But there is a twist: Don't be misled by that arc-like motion. If we plot our changing position with respect to *time* during this portion of the ride, we get a periodic sine wave. The fact that the pendulum swings in predictable periods is why it became the basis for the pendulum clock.

There is another relevant curve called the Witch of Agnesi, named after eighteenth-century mathematician Maria Gaetana Agnesi. The eldest of twenty-one children,* Agnesi was known in her family as the Walking Polyglot because she could speak French, Italian, Greek, Hebrew, Spanish, German, and Latin by the time she was thirteen. Agnesi had the advantage of a wealthy upbringing; the family fortune came from the silk trade. And she also had a highly supportive father, who hired the very best tutors for his talented eldest daughter and insisted she participate in regular intellectual salons he hosted for great thinkers hailing from all over Europe.

The young Maria delivered an oration in defense of higher education for women in Latin at the age of nine; she translated it from the Italian herself and memorized the text. Contemporary accounts suggest that Agnesi loathed being put on display, even though her erudition earned her much admiration. One contemporary, Charles de Brosses, recalled, "She told me that she was very sorry that the visit had taken the form of a thesis defence, and that she did not like to speak publicly of such things, where for every one that was amused, twenty were bored to death."

De Brosses admired her intellectual prowess greatly, and was horrified upon learning that she wished to become a nun. She did become a nun, but not before spending ten years writing a seminal mathematics textbook, *Analytical Institutions*, published in 1748—the first surviving mathematical treatise written by a woman. She was also the first woman to be appointed a mathematics professor at a university (the University of Bologna), although there is no record she ever formally ac-

* I was relieved to learn her father married three times, since the thought of one woman enduring that many pregnancies boggles the mind.

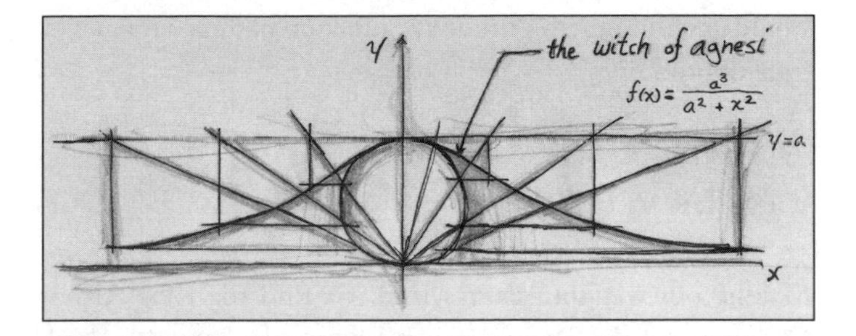

cepted the position. She died a pauper in 1799, having given away everything she owned.

But her work lives on. One of the curves featured in *Analytical Institutions* is the Witch of Agnesi. Agnesi dubbed it *la versiera*, a nautical term meaning "a rope that turns a sail"—an allusion to the motion by which the curve is drawn.

At some point, a harried English translator misinterpreted the word as *l'avversiera*, "she-devil" or "witch."

What does this have to do with the pendulum motion of the Devil Dive? Among other things, this curve describes a driven oscillator near resonance—a swinging pendulum that is being poked or prodded to keep it in motion, for example, like someone pushing a child on a swing. When the rate of prodding matches the rate of the pendulum's swing, it is said to be in resonance. If the rate of prodding is very, very close to the rate of the swing, the amplitude (height) of the swing, plotted as a function of frequency, forms the Witch of Agnesi. We've already seen that the physical motion of a pendulum forms an arc, while plotting its position as a function of time gives us a periodic sine wave. So had someone (or something) been pushing us during the pendulum phase of the ride at almost the exact same rate as our swing, the Witch of Agnesi would describe our amplitude as a function of forced frequency (rate of prodding).

V IS FOR VECTOR

Making our way into Fantasyland, we find the King Arthur Carousel and the *Dumbo*-inspired flying-elephant ride, both excellent examples of rotation around a fixed axis. But it is the Mad Tea Party—usually called the spinning teacup ride—that provides us with an unusual illustration of vectors: motion in specific directions. A vector is technically defined as any quantity having both direction and magnitude. In physics, vectors typically describe force, velocity, acceleration, or similar three-dimensional properties. How the different vectors combine determines their net strength; one must take into account not just how strong a given force might be, but also in what direction it is pushing.

It is easiest to illustrate the concept in one dimension. Picture your standard number line. An object moving in a straight line has a direction, depicted by a small arrow above the number. If it starts at 0 and ends at 5, this is called vector (5); it's the same as any other number along that line, except we have specified a direction. Because it's moving left to right, it is a positive number. A vector pointing from right to left would be a negative number. Vectors can be added together or subtracted, just like regular numbers. Combine vector (5) with vector (–5), and the two cancel each other out completely; combine it with vector (–3), and you end up with vector (2); combine it with vector (4), and you end up with vector (9). And so on.

Frankly, vectors aren't very interesting in the one-dimensional realm of the number line: There is no real difference between them and ordinary numbers. In two dimensions, vectors are pairs of numbers (Cartesian coordinates) that describe the direction of movement in a plane. In three dimensions, they describe directional motion through space using three coordinates.

Here is how vectors apply to the Mad Tea Party. Any rotating body's motion has a vector that is constantly changing, because the direction shifts at each point in the turn. The teacup ride in Disneyland consists of a series of rotating circles, or turntables, each moving along its own vector that is constantly changing its direction because of the rotational motion. There is one big circular moving platform that rotates clockwise. Within that circle are three smaller ones that rotate independently, counterclockwise, and within each of *those* circles are individual teacups that rotate clockwise, independently from the two bigger circles.

The riders can spin their teacups as fast as they want by turning the metal wheel at the center of the cup, applying a torque to increase the teacup's angular momentum, and hence the rate of spin. As Sean and I strain mightily to spin our teacup as fast as possible, I notice something intriguing. Every now and then we achieve an especially sharp, fast rotation, whereas at other points, no matter how hard we pull that metal wheel, we can't achieve much rotation at all. Sean explains that this is because of dueling vectors. Sometimes the vectors work against each other, pulling in different directions and canceling each other out, to varying degrees. At other times in the rotation, they add together, all pulling in the *same* direction, so we spin that much faster.

Space Mountain—Tomorrowland's main attraction—provides us with the quintessential example of a calculus problem

involving vectors. When Walt Disney first designed Tomorrowland, he noted that it would be out of date almost immediately. By twenty-first-century standards, the "future" it envisions is downright quaint, harking back to a more innocent era. Tomorrowland didn't even *have* a roller coaster until Space Mountain opened in May 1977, after the original ride proved so popular at Disney World. Disney didn't live to see it completed. Space Mountain took two years to build and cost upwards of $20 million, and the park set an attendance record the first weekend the ride opened. Six of the original seven Project Mercury astronauts were on hand for its inauguration.*

In the 1968 film *2001: A Space Odyssey*, astronaut Dave Bowman (played by Keir Dullea) walks down a long white

* Scott Carpenter, Gordon Cooper, John Glenn, Wally Schirra, Alan Shepard, and Deke Slayton. Alas, Gus Grissom was one of three astronauts killed ten years earlier in a tragic launch-pad fire.

circular tunnel to the space ship that will carry him on his mysterious mission into deep space. It's difficult not to recall Kubrick's masterpiece while waiting in the long line for Space Mountain. The ride's interior is eerily similar in design. We follow winding metal ramps down into the bowels of the coaster, encountering the occasional video screen showing famous astronauts talking about their missions. Finally, we reach the front of the line and take a seat inside our little rocket-shaped car.

We rise to the top of first one, then another lift hill, winding through a passage that features glowing red bars that seem to be rotating. At the top of the third and final lift hill, our rocket pauses briefly as we gaze out into the vast darkness of "space"—there appear to be thousands of stars and galaxies, when in fact it is simply a clever effect achieved with mirror balls scattered throughout the ride's interior. A voice announces, "You are go for launch," and pure gravity takes over as our rocket begins its rapid descent, accelerating through the remainder of the track. The sensation is enhanced by gusts of wind from strategically placed air vents as we careen and lurch through the darkness. When it is time for our "reentry," we decelerate and return to the docking station.

For all its futuristic trappings, Space Mountain is a classic roller coaster, from a physics standpoint.* Roller coasters operate on inertia, gravity, and acceleration—and the greatest of

---

* Which is not to say there isn't considerable art involved in designing a good roller coaster. "This isn't rocket science; it may be more complicated than that," Space Mountain's ride track engineer Bill Watkins recalled. "Once a rocket leaves the Earth's atmosphere, there is little drag to contend with . . . [and] they don't have to worry about getting a Mickey Mouse hat caught in their wheels."

these is gravity. Our rocket builds up a large reservoir of potential energy while being towed up those three initial lift hills. The higher we rise, the greater the distance gravity must pull it back down, and the greater the resulting speeds. As our rocket starts down the first hill, all that accumulated potential energy is converted into kinetic energy and our car speeds up, building up enough kinetic energy by the time it reaches the bottom to overcome gravity's pull and propel the car up the next hill. And so on for the rest of the ride.

Sean likes Space Mountain. A lot. Apart from the obvious fun factor, he declares that we can use calculus to determine our trajectory (the path we took) when all we know is our acceleration. Space Mountain has none of the intricate maneuvers that have become standard among more extreme coasters: fancy corkscrews, loops, and so forth. Instead, it relies on a series of shorter dips and sharp turns in near-total darkness. Because we can't see the track, we can't anticipate where we are likely to go next or prepare for the sudden shifts in velocity. The few visual cues we are given are deliberately misleading.

We can still figure out which path we took, because we can feel the physical effects of acceleration on our bodies and deduce our trajectory from that data. These are the g forces that describe how much force the rider is actually feeling; *g* is a unit for measuring acceleration in terms of gravity. Our rocket is constantly accelerating over the course of the ride: forward and backward, up and down, and side to side. Our inertia is separate from that of our rocket, so when it speeds up, we feel pressed back against the seat because it's pushing us forward, accelerating our motion. When the rocket slows down, our bodies continue forward at the same speed in the same direction, but the restraining bar decelerates us to slow us down. All this acceleration produces corresponding variations in the apparent strength of gravity's pull. For example, 1 g is the force of Earth's gravity: what the rider feels when the car is stationary or moving at a constant speed. Acceleration causes a corresponding increase in weight, so that at 4 g's you will experience a force equal to four times your weight.

That gives us an intuitive sense of our trajectory throughout the ride, but for a truly rigorous analysis, we should have had the foresight to bring along a makeshift accelerometer. As electronic components have continued to shrink, accelerometers became easier to embed. Our matching his-and-hers iPhones come with built-in accelerometers, which is how the device knows when to adjust the screen from a vertical to a horizontal view when you turn the phone on its side. If our little rocket came equipped with a built-in accelerometer—yes, there is an app for that—that accumulation of data would give us our acceleration function.

Let's start with a simplified version of this standard textbook problem, assuming that we are moving in a perfectly straight line. How can we figure out our trajectory—our position as a function of time—knowing just our acceleration? Our acceleration accumulates over time to give our velocity, Sean explains; we accumulate our increasing speed at each moment in time to determine our final velocity. So that means velocity is the integral of acceleration. Velocity in turn increases over time to give position, so position is the integral of velocity. "You just have to integrate the acceleration twice to figure out position as a function of time," Sean concludes triumphantly—just like our free-fall problem.

However, there is a complicating factor: The rockets move left and right and up and down, not just forward in a straight line. So not only are the rocket sleds constantly shifting between potential and kinetic energy, but every time we shift direction, we also are shifting vectors—our direction of movement is constantly changing. Thanks to an unjustly obscure nineteenth-century British mathematician and physicist named Oliver Heaviside, we have the tool to solve this complex conundrum: vector calculus.

A product of the London slums that also produced Charles Dickens, the red-haired, diminutive Heaviside fell ill with scarlet fever as a child, which left him partially deaf. His social skills seem to have suffered as a result: He didn't get along with the other children at school in Camden Town, although he was a top student in every subject save geometry. Perhaps traditional education couldn't contain his eccentric genius: He dropped out at sixteen to continue his schooling at home. It helped that his uncle was Sir Charles Wheatstone, who coinvented the telegraph in the 1830s and was a recognized expert in the new field of electromagnetism. Within two years, young Oliver found himself working as a telegraph operator, quickly advancing to chief operator. It was the only full-time employment he ever experienced.

One could blame James Clerk Maxwell, the prominent physicist who first formulated the set of equations for electromagnetism that still bear his name, for Heaviside's sudden shift into the ranks of the chronically unemployed. Heaviside discovered Maxwell's seminal treatise in 1873 and was so enthralled by the work that he quit his job the following year to study it full-time, moving back into his parents' home in

London. (History has not recorded his parents' reaction.) Once he'd grasped the essential points, "I set Maxwell aside and followed my own course," Heaviside later recalled. In the end, he reduced Maxwell's equations from twenty down to four vector equations and built upon that work to develop vector calculus.

Few objects move in a straight, flat line. We don't drive down straight roads with no turns or hills, and a roller coaster would be a very dull ride indeed if it only moved in flat, linear motion. Vector calculus lets us solve the same calculus problem in three dimensions: retracing our path by determining our position at each instant over the course of the ride. We describe position in three-dimensional space with three Cartesian coordinates (x, y, and z), so there are now three numbers involved in our calculations. Nothing else has changed from the previous example: Our trajectory is still position as a function of time, and thanks to the data gathered by our accelerometer, we know our acceleration as a function of time. Acceleration builds up to give us our velocity, which in turn builds up to give us our position at any moment. It's just more complicated because our movement has a constantly changing direction. We must keep track of three directions at once, and each has its separate position, velocity, and acceleration.

Heaviside never gained the recognition he deserved until after his death in 1925; he was justly bitter about this. He became quite eccentric in his later years, spending the last two decades of his life as a virtual recluse in Torquay, near Devon. He suffered bouts of jaundice and the tormenting of neighborhood children, who threw stones at his window and scrawled graffiti on his front gate. Neighbors reported that his home was furnished primarily with huge granite blocks. Otherwise scruffy and unkempt, he took to painting his impeccably manicured fingernails bright pink, and signing letters with the mysterious initials W.O.R.M. after his name—providing ample fodder for future armchair psychoanalysts as to what the letters might have meant to him. Perhaps he would have found comfort in the fact that, over a hundred years later, his method of vector calculus would one day shed light on a budding calculus student's encounters with the rides at Disneyland.

MAKING A SPLASH

By late afternoon, we have worked our way over to Critter Country, where the skyline is dominated by the soaring peak of Splash Mountain. Splash Mountain is a "log flume" ride. Loggers used to transport logs down mountains to a sawmill by floating them down the river. Eventually someone had the brilliant idea of hollowing out those logs and using them as makeshift boats. The first artificial log-flume ride—called El Aserradero (The Sawmill)—opened in 1963 at the Six Flags theme park in Arlington, Texas. The Disneyland version is a large plaster mountain housing canals (or flumes) filled with water, and artificial hollow logs that can seat up to six people. The flow of water along the flumes propels the log boats forward, with a little help from mechanical chains and pulleys to hoist the logs up the hills. Just as with a roller coaster, good old-fashioned gravity does the rest.

Nothing says Disney like plaster facades and cheesy animatronics. This ride takes its inspiration from *Song of the South*, with scenes depicting the adventures of Br'er Rabbit. The robotic critters* lining the "banks" of the faux canal snaking through Splash Mountain serenade us as we float along, with a jaw-clenching ditty about positive thinking, finding your "laughing place," and having a zip-a-dee-doo-dah day! Just as I am wishing I had a stun gun capable of overloading their circuitry with an electromagnetic pulse, we come to a sudden drop and plunge into the depths of the cavernous "briar patch."

SPLASH! The front of the canoe hits the bottom and displaces a large amount of water. Sean is drenched from head to toe, and instantly regrets his chivalrous offer to take the front seat instead of me. Nor does the dousing end there. We soon experience another sudden drop with accompanying splash, and another, and then must endure the shrieking laughter of the animatronic animals reveling in our plight. They have gone from abrasively cheery to vaguely sinister; we even spot Br'er Rabbit on the bank, tied up and struggling, about to be eaten by Br'er Fox. And those mechanical vultures with glowing red eyes look eager to gnaw with abandon on our sodden bones. The animals have found their laughing place, and it is called Das Haus von Schadenfreude.

There is one last lift and one final, fifty-foot drop, accompanied by yet another dousing. This is one of the fastest rates of descent in the entire park. While we know from our exercise with free-fall rides that the collective weight of everyone in our log does not affect the rate at which we fall, it *does* help determine how wet we are likely to get on this final splash, because

* Many of the animatronic animals are recycled from an older, less popular attraction called "America Sings," which closed in April 1988, because construction of Splash Mountain was already far over its $75 million budget. Sadly, the animals still sing.

the amount of water displaced is proportional to that collective weight.

The good news is that despite being drenched, our log boat floated and didn't sink, because our average density was less than that of water. Had we sunk, we would have faced a dilemma reminiscent of our old friend Archimedes.

When not drawing countless rectangles under curves, he was having spontaneous epiphanies in his bathtub. Legend has it that Archimedes accepted a challenge from a local tyrant, Hiero of Syracuse. Tyrants are not trusting by nature, and Hiero was no exception. He was convinced a local goldsmith he had hired to make a golden wreath as a gift to the gods had cheated, replacing some of the gold with silver. No self-respecting deity would accept a cheap alloy. But how could he prove dishonesty? Hiero turned to Archimedes for help, who promptly went to the public baths for a good long think. He noticed that the more his body sank into the water, the more water was displaced.

The weight of an object pushes water out of the way, Archimedes reasoned, and the water in turn pushes back. So the buoyant force exerted by a fluid, like water, is equivalent to the weight of the fluid displaced. This gave him an idea for how to test the golden wreath: Gold weighs more than silver, so a crown mixed with silver would need more bulk to achieve the same weight as a crown made of purest gold. He could weigh the crown and submerge it in water to measure its volume, and from that he could calculate the density. Archimedes had stumbled on a way to calculate the volume of irregular objects very precisely. Euphoric over this critical insight, he leaped out of the tub and ran stark naked into the street, shouting "Eureka! Eureka!"* Once he determined the crown's volume, then the ratio between its weight and its volume would indicate its density and answer Hiero's question of purity.

So let's imagine that, instead of floating, the log boat sank with all its passengers. We can ask everybody to hold their breath while we use Archimedes' principle to determine the total volume and from that, to calculate their average density. But even had I convinced Disneyland (and my fellow passengers) to let me do that experiment despite the liability issues, I would still lack another crucial piece of the puzzle—I had failed to note the weight of all the other passengers. This is an object lesson in why it's so important to carefully collect one's raw data while doing the experiment.

If we know the combined weight of the passengers and the log we are riding in, the volume of our log, and the collective density (in units of grams per cubic centimeters), we can divide the total weight by the total density to get our volume in cubic meters. We also need to know the density of water; a quick Google search reveals that one liter of water has a density of 1 kilogram. Now we multiply the volume of our log and its passengers by the density of the water to find the volume of water displaced. Those hollow plastic logs hold six riders of varying weights. Assuming an average weight of 150 pounds per passenger (150 × 6, plus the weight of the log itself), that gives us a pretty substantial volume—and a substantial displace-

* Eureka (Greek *heurēka*) means "I've found (it)," and ever since, surprising scientific insights have been known as eureka moments. Ironically, Archimedes most likely never said that, certainly not while running naked through the streets. Blame the Roman architect Vitruvius, who first recorded the anecdote two hundred years after Archimedes' death.

ment of water when we hit the bottom of that final plunge. No wonder we're completely soaked by the ride's end.

Sodden jeans and sneakers are not a pleasant sensation. It is a cool, cloudy day for Anaheim and late enough in the afternoon that our clothing takes longer to dry than it would on a warmer day. While we wait, Sean explains that there is a calculus problem in our current plight: The rate at which our clothing dries—that is, the rate of evaporation of water from the fabric—forms an exponential decay curve. It is similar to the rate at which a cup of hot coffee cools until it reaches thermal equilibrium with its surroundings.

The coffee cools off very quickly at first, but as it gets closer to thermal equilibrium, that rate of cooling slows down and eventually levels off. This is because the amount of heat lost is proportional to the temperature of the coffee: It is determined by the ratio of the excess heat to the lower temperature limit—how cool the coffee can get, usually ambient room temperature. So as the coffee cools down and gets close to room temperature, there is less excess heat and thus a smaller ratio between the two variables. And the rate of cooling levels off.

The same thing happens with the evaporation of the moisture in our clothing. Plot the rate of evaporation as a function of time, and you can see this in the resulting curve: There is a steep drop initially, followed by a gradual leveling off. The alert reader will note that because we are dealing with a rate of change, we must be taking a derivative. That means we can find an answer to the question, "How fast is the water in our clothes evaporating at x time?"—a form of the velocity function, similar to determining our instantaneous speed in chapter 2—by finding the slope of the tangent line along that particular point on our curve.

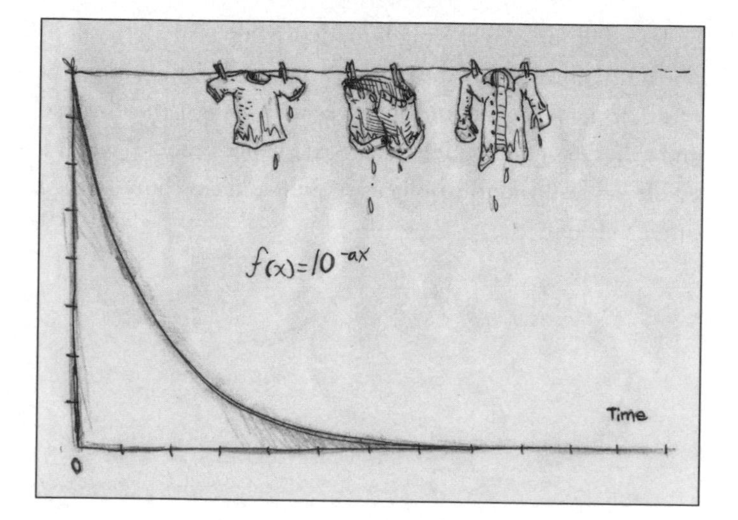

We experience this exponential decay curve firsthand and soon find ourselves wondering, *Will we forever be slightly damp?* It is beginning to feel that way, and we have dinner reservations in an hour at the Blue Bayou restaurant—the sole fine dining establishment in Disneyland, situated just inside the Pirates of the Caribbean ride. We end up squishing our way over to the gift shops in New Orleans Square in search of a change of clothes, where the sales clerk assures us this happens all the time. They do a brisk business, thanks to Splash Mountain.

And thus we find ourselves, an hour or so later, seated in the Blue Bayou's fake outdoor grotto in matching Pirates of the Caribbean hooded sweatshirts, my outfit completed by a jaunty newsboy cap with a skull-and-crossbones motif to hide my hopelessly tangled hair. By this time Sean is very much in need of a drink, and the Tinkerbell Fruit Punch with a fairy light garnish simply isn't going to cut it. Alas, there is no alcohol to be found in the Magic Kingdom, depriving us of a prime opportunity to work out the calculus of inebriation. (Oh yes, it can be done.) We content ourselves with a sugar rush instead and split the signature dessert: a boat-shaped "cookie" with an edible sail featuring the obligatory skull and crossbones. It is a pirate's life, indeed.

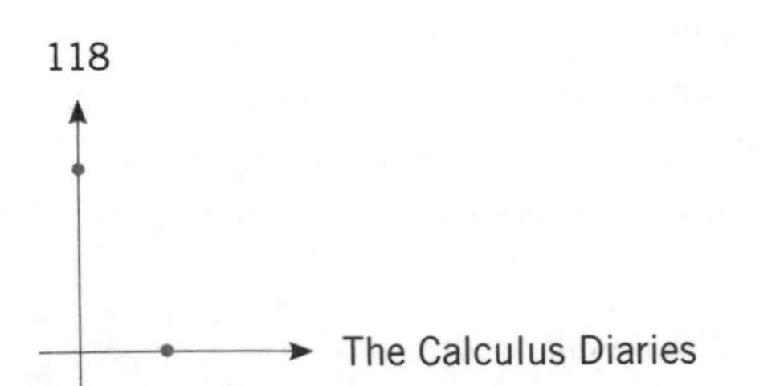

5

Show Me the Money

> It is clear that economics, if it is to be a science at all, must be a mathematical science . . . simply because it deals with quantities. . . . As the complete theory of almost every other science involves the use of calculus, so we cannot have a true theory of economics without its aid.
>
> —W. S. JEVONS

Like beauty, an object's intrinsic value rests in the eyes of the beholder. One man's priceless treasure is another man's culinary delight. In seventeenth-century Holland, a hungry sailor mistook a rare tulip bulb that was on display for an onion and stole it from a local merchant. The merchant chased him down Amsterdam's busy streets, catching up just in time to find the sailor "eating a breakfast whose cost might have regaled a whole ship's crew for a twelvemonth." That was the going rate for a single bulb at the height of what is now called tulip mania. Incensed, the merchant had the sailor thrown into prison for his crime.

That is one of the more outlandish anecdotes about tulip mania popularized in the nineteenth century with the publication of Charles Mackay's *Extraordinary Popular Delusions and the Madness of Crowds*. Today, the excesses of tulip mania are the stuff of legend, trotted out as a cautionary tale whenever economists find themselves analyzing catastrophic bubble markets.* Modern economists dispute many details of Mackay's account, which was based on dubious source material, but it makes for lively reading. And while the sailor's story might not be true, it epitomizes the kind of irrational exuberance and frenzied overvaluation of assets that so often serve as harbingers of economic disaster.

How did the tulip become such a collector's item in the first place? Holland is widely known as the land of colorful tulips, and one would think the bulbs would be a cheap commodity. But the bright bell-shaped flower is actually a relative newcomer to the country. In 1593, a Dutch botanist named Carolus Clusius returned from a trip to Constantinople with a few precious tulip bulbs and planted them in his garden, supposedly to study them for medicinal purposes. Then his neighbors broke into the garden and stole some of the bulbs, figuring—correctly—that the exotic flora would bring in a pretty penny. Thus was born the Dutch tulip trade and the

---

* There is still considerable debate as to whether tulip mania constituted a true bubble market in modern economic terms. A bubble forms when investors place so much demand on a product that the price soars far beyond what that product could possibly be worth. Wikipedia offers a corollary to that definition: "For tulip mania to have qualified as an economic bubble, the price of tulip bulbs would need to have become unhinged from the intrinsic value of the bulbs." Did this happen or not? Discuss.

onset of a collective mania that drove prices to dizzying heights.

One recorded list of items traded for a single tulip bulb included a bed, some clothing, and a thousand pounds of cheese, but prices rapidly escalated beyond such humble items. In 1624, a buyer offered 3,000 guilders (equivalent to a year's earnings) to a man in Amsterdam in exchange for a dozen specimens of the rarest tulip, known as Semper Augustus and identifiable by its blue-black petals accented with streaks of crimson and a sprinkling of white. A sale of forty bulbs for 100,000 guilders was recorded in 1635. The most expensive bulbs were far too valuable to be planted, so instead it became the fashion among their (once) wealthy owners to display the plain bulbs—well away from the gaze of famished sailors.

Speculators were desperate to cash in on the gravy train, mortgaging whatever they could to raise capital to invest in a few "starter bulbs," in hopes of jump-starting a lucrative business in the tulip trade. One transaction records the trade of a farmhouse in 1633 in exchange for three rare bulbs. There was even a thriving futures market for tulip bulbs, with business often being conducted in local taverns; at the height of the frenzy, one bulb changed hands ten times in a single day. But the tulip bubble burst almost as quickly as it formed. One day a buyer didn't show up with the cash, and panic set in and spread. Within days, bulbs that had sold for staggering sums were now "worth" roughly one-hundredth of their former value.

Such are the harsh realities of supply and demand. Those Dutch tulip speculators might have benefited from a spot of calculus. (Unfortunately it hadn't been invented yet.) The tools of calculus are particularly well suited to the financial sector, which deals heavily in rates of change: inflation, interest rates, mortgage rates, and the impact of supply and demand on pricing can all be described by functions linking one feature to another. We can use the derivative to determine the rate at which one factor changes relative to another, and we can employ the integral to determine the cumulative effect of any ongoing process. In the case of tulips—or any product, for that matter—supply and demand are interdependent quantities: A change in one affects the other, and choices about production and supply affect the profit you make. The integral comes into play when calculating interest, whether accumulating interest on a savings or retirement account or calculating the interest on a mortgage loan.

TIPTOE THROUGH THE TULIPS

Why did the tulip market go boom, then bust? There were several contributing factors, but it had mostly to do with simple supply and demand. The tulip bulb was a rare commodity from the start, although ordinary bulbs were often sold by the pound. Then some of the tulips contracted a mosaic virus that altered the color of the blooms, streaking their petals with scarlet. Those varieties were even more rare, attracting wealthy collectors and commanding an even higher price. Demand grew so rapidly that the supply of bulbs could not keep pace, and prices rose and rose.

Dutch residents were flush with extra cash after the end of hostilities with Spain. Amsterdam's merchants were thriving at the center of the lucrative East Indies trade, earning profit margins as high as 400 percent in a single voyage. So the market could absorb—temporarily—the outrageously high prices demanded for tulip bulbs. But no market can sustain that kind of exponential growth rate indefinitely. Eventually the price became so high that very few buyers were able to meet it. Once that first buyer didn't show up for the sale, a domino effect occurred. Demand dropped suddenly, panic ensued, and the bubble burst, with dire economic consequences for those who had speculated on the market.

Let's imagine that I am a tulip dealer in seventeenth-century Holland, eager to turn a tidy profit in this burgeoning industry. I am drawn to tulip bulbs because they command a hefty price and there are still a substantial number of buyers willing to pay that price. Also, flowers are pretty. I just have to be careful not to raise the price so much that I chase away prospective buyers; if prices get too high, demand will drop, and my profits will never materialize. Ideally, I want to maximize my profit—which will be the gross revenue I bring in with the sale of my exotic tulip bulbs, less the associated costs I incur to obtain them—and minimize my production costs. Calculus can help me do this.

The cost of producing a given product depends on how many items are produced. If I decide to print flyers advertising my tulip bulbs, there is a basic cost I will incur for setting up the equipment to do so. It's probably not worth that initial cash outlay to print only a hundred flyers; I'm better off printing twelve thousand flyers and stocking up for the future. Or am I? There might be storage costs to consider, and these must be offset against the money I save by printing more flyers. Perhaps it would be better to make two print runs of six thousand flyers each. I need to strike just the right balance between these two factors.

Assume I have fixed setup costs of $2,000 for the printing press. The cost of storing twelve thousand flyers is minimal—$3 per year—but I still need to factor that into my financial planning. With these two bits of information, I can devise an equation that gives me the total cost of maintaining inventory plus the produced and setup costs. I designate y as the number of print runs, and each run costs $2,000. The number of flyers produced and stored is represented by x. But it's not going to be $3 constantly; x fluctuates over time, unlike y, which is fixed. My storage space is full after every production run, but as I hand out flyers over time, the number in storage steadily decreases, until all the flyers are gone and my storage costs are back to zero. So I take the average storage cost, which will be half of $3: $1.50.*

I end up with a total cost of ($2,000$y$) + 1.5$x$. Multiply x and y—the number of flyers I produce with each print run times the number of print runs—and I get the total number of flyers printed over the course of one year: 12,000. I can simplify my equation by eliminating y entirely, because it is equivalent to 12,000 over x. This means I can rewrite the total cost equation as $2,000 times 12,000 over x, plus 1.5x, to get my "cost function," and once I have that, it's a relatively straightforward process to determine how often I should order a print run. I just need to minimize the sum of the storage costs plus the setup costs. I can find that "sweet spot" on the graph by setting the derivative of the cost function equal to 0 and then

* I am cheating a little by assuming a constant rate of change.

figuring out what value of x gives that answer. In this case, my best bet would be to make three print runs of 4,000 flyers each over the course of a year.

Now we estimate the expected revenue based on how much of a product I produce. How do I price my tulip bulbs in order to maximize my profit? Tulip bulbs incur a lot of initial costs, unless I opt for the sneaky alternative of stealing them from my globe-trotting neighbor, Carolus Clusius. Even then, my theft would yield a very limited supply. But it might bring in sufficient revenue to finance my little start-up venture. It takes about seven years to grow tulips from seeds: There would be costs associated with renting a greenhouse, buying fertilizer, watering the seeds, and so forth, over the seven-year incubation period for producing the tulip bulbs. And each bulb can produce only a few clones before expiring, so there will always be a limited supply of bulbs. (Only bulbs produce genetically identical offspring; seeds introduce genetic variability.)

Let's assume that my fixed cost will be \$100,000 and that it costs around \$30 per bulb on top of that to "make" my product (the bulbs). So my function for cost is \$100,000 + $30q$, where q stands for the quantity of bulbs. The change in cost is called the marginal cost; it measures the incremental expense of producing one more tulip bulb. Then there is the marginal revenue, the rate at which the revenue increases with the production of one extra bulb—in other words, it's a derivative.

Starting with an estimated production of 20,000 bulbs, I can determine a maximum and minimum price (p), where at a given price, approximately $20{,}000 - 50p$ bulbs will be sold. At a maximum price of \$400, there would be no buyers, and if I gave away the bulbs for free, all 20,000 bulbs would find a home with a buyer—if someone who pays nothing can be described as a buyer. If I sold them for $100, however, 15,000 bulbs would be sold, according to my spiffy formula (100 × 50 = 5,000, which we then subtract from the 20,000 total bulbs). So my revenue R is equivalent to the price per bulb multiplied by the number of bulbs I sell, or $1.5 million.

We want to set the marginal cost equal to the marginal revenue. That's where the maximum profit will be. If the marginal revenue is greater than the marginal cost at a particular production level, then growing one more tulip means the increase in revenue will be greater than the increase in cost, and I make more profit. If the marginal cost is greater than the marginal revenue, I will also increase my profit, this time by growing fewer bulbs, because I will reduce my costs more than I will reduce my revenue. The answer: I should grow 9,250 bulbs and sell them at $215 each in order to maximize my profits.

That's roughly how the market should work under ideal conditions, but we do not live in a simple world. Something has gone seriously amiss when a rare tulip bulb possesses more value than a farmhouse. The exponential decay curve decreases rapidly initially and then gradually slows its rate of change; the exponential growth curve exhibits similar behavior in reverse. But when a bubble forms, the result is a so-called boom-and-bust curve: Growth starts out increasing exponentially but peaks and collapses quite suddenly. Those who enter a hot new market early may reap enormous profits, but as more and more people enter the fray over time and prices go up and up, there are fewer and fewer buyers. Eventually the market will hit a peak and collapse—and the "decay" will be steep and sudden. That's what happened with tulip mania. What happens when the bubble mentality comes to real estate and literally hits people where they live?

HOME SWEET HOME

It's a bit disheartening to tour a foreclosed home; a sense of loss seems to permeate the space. Despite being only four years old, the town house we are touring has seen better days: The floors are scuffed, the window screens are torn, and the previous owners have absconded with the appliances as compensation for losing their home. While the unit is spacious, the interior feels cramped and dingy on this overcast afternoon, particularly since the electricity has been turned off. But the building is in a prime location, a few blocks from many of our favorite shops and restaurants.

Two years after moving to Los Angeles, we have joined the ranks of nervous house hunters, cautiously dipping a toe into the volatile Southern California real estate market to test the waters. We are in no hurry. Our rental apartment is sufficient for the short term: It's in a very walkable location in downtown Los Angeles, with free parking in the garage across the street, and a friendly full-time concierge named Mike. But we are running out of space, having merged two households when we got married. Most of our books are in storage. The dining-room table is strewn with Sean's books and physics papers, while the second bedroom performs double duty as my office and a cramped guest room. And there isn't nearly enough closet space.

It is both the best and worst of times to buy. We have been patiently waiting for prices to drop to more affordable levels, and in the wake of the economic collapse of September 2008, housing values are plummeting. The median home price in California dropped 41 percent—more than double the 16 percent decline in median home prices for the United States as a whole—between February 2008 and February 2009, according to the California Association of Realtors, as a tidal wave of foreclosures drove down values. Nobody knows how much farther prices will fall, and that uncertainty means everyone is a wee bit skittish, particularly the banks: Loans are much harder to come by, even for highly qualified buyers. The process is fraught with anxiety—starting with the search for a suitable home.

Every prospective home buyer knows there is no such thing as the perfect place. The exercise of touring several homes helps us get a sense of the market, what we can afford, and the factors that matter most to us. We know we need three bedrooms (or two bedrooms with a den), with parking for two cars. We prefer central locations with shops and restaurants within easy walking distance. Such areas tend to have higher housing costs, so we know we will have to make a trade-off between square footage and prime location. We like to entertain, so a spacious living area with open kitchen is desirable. And we don't want to do any heavy remodeling. Can we find the optimal combination of our desired features—within our price range?

One unit has an awkward layout. Another boasts a dramatic curving staircase in the foyer, but there is no extra space for an office—a prime consideration for a professional writer. I like a faux-Spanish townhouse, but it is less to the taste of my more modernist spouse. One home has bizarre blue plastic kitchen cabinetry; in another, the bathtubs are freakishly small for my six-foot-one-inch spouse; and yet another suffers from cheap flooring and astronomical home association fees. All are preferable to the dingy "penthouse" unit with stained carpets, chipped tiles, and a "rooftop deck" lined with sticky tar paper.

As for that first battered foreclosure, we decide the living/dining area is too cramped for our needs.

House hunting is the ultimate experiment in comparison shopping: weighing different variables and seeking the optimal combination of those factors. In a sense, we are doing conceptual calculus. Mathematicians merely take this process to the next level by quantifying everything and organizing that data into an equation. In principle, we can turn our house-hunting experience into a multivariable optimization problem, similar to what we did to determine the optimal price for our tulip bulbs in order to maximize profit. We just need to find some way to quantify our subjective criteria.

Because we need a continuous curve, we'll assume we have an infinite number of houses to choose from. Anyone who has undertaken serious house hunting knows it can feel like infinity sometimes. Calculus will help us narrow the search by optimizing our happiness with our final choice. For simplicity's sake, we will restrict our variables to two easily quantifiable qualities: square footage (q) and walkability (w), the latter based on an online "walkability score" algorithm. That gives us our function: $f(w,q)$. The "curve" for this will look much different from the graph for a function with a single variable: It will be a contoured surface floating above a plane.

Think of a map that only shows your location with two intersecting points: latitude and longitude, or the place where Wilshire Boulevard meets Figueroa Street in downtown Los Angeles. What is lacking is the altitude. Bringing in a second variable to our optimization problem is like adding altitude to a map, so we can tell not just where we are, but the elevation of that particular spot. Not only do we have the x and y axes on our Cartesian grid—representing walkability and square

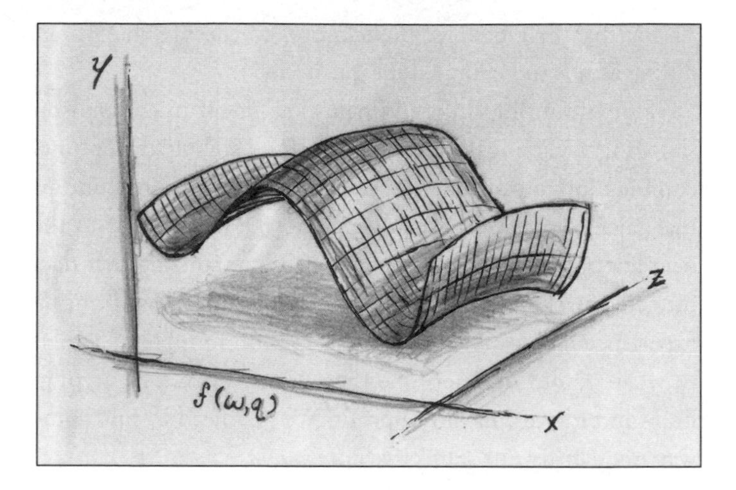

footage, respectively—we also have a third, the z axis, jutting out at an angle.

If we merely consider square footage and walkability, what stops us from increasing those two variables to infinity to gain optimal happiness? Clearly we need some kind of constraint, and we find it in the price. We do not have an infinite amount of funds, so we need to build a third aspect into our "happiness function": cost. We can assume that cost depends directly on size and walkability. One of the first steps prospective home buyers take is determining their price range. Go beyond that price range, and our happiness will start to decrease again, even though square footage and walkability continue to increase. If we can't afford it, we won't be as happy.

We plot happiness as a function of our two variables (w,q) to get a nice smooth curvy plane that goes up, peaks, and descends after the peak. Then it is simply a matter of taking a derivative of each variable separately—this is called partial differentiation, or taking a partial derivative—and finding the value that sends both to zero. That will be the point(s) on our curve where the slope of the tangent plane is zero (horizontal). Wherever the tangent planes are flat is where we will find our optimal solution. That is where we will find maximum happiness with our choice. We find we must indeed make a trade-off between walkability and square footage. The price per square foot is significantly higher in very walkable locations, so we can't afford as much square footage in prime areas and still stay near the peak of our happiness curve. Similarly, beyond a certain point, too little square footage will also decrease our happiness. Finding the "sweet spot" on our multivariable curved surface enables us to narrow our options down from infinity to three:

Option 1: This is a three-bedroom, three-bath "architectural" townhouse featuring bamboo floors and cabinetry, and a wall of windows bathing the main loft area in sunlight. There are ample closets and a private two-car garage. The location isn't as walkable as we would like, but the price per square foot is below market rate, so we would get a lot of space for the money.

Option 2: This is a three-bedroom, three-bath condominium. The interior features dark woods, and Asian influences abound. It is slightly smaller, but there are many closets, and there's a large balcony off the dining room. The drawbacks are the tandem parking spots in the communal garage (side by side is preferable) and the location, which is not very walkable.

Option 3: This is a two-bedroom, two-bath unit with den in a prime location with excellent walkability. The design features Mediterranean influences, with a balcony, spacious living room and open kitchen, and luxurious baths. The price per square foot is significantly higher, so we can only afford one of the smaller units. There are fewer closets, and the parking spaces in the shared garage are tandem.

All else being equal, how do we find the optimal choice among them? Calculus is less helpful here. Ultimately, one's choice of home is an emotional, subjective decision. But we do engage in an approximation of an optimization problem whenever we comparison shop; it's one way to bring some rationality to the process. Yet even then, our choice of how much to weight a given variable is highly subjective. Dutch psychologist Ap Dijksterhuis studies how house hunters are often subject to "weighting mistakes." Given the choice between a larger home in the suburbs with a longer commute, and a smaller, more expensive home in a central location, most home buyers opt for the larger home. They underestimate the negative impact of a long commute on overall quality of life over time.

DOWN THE RABBIT HOLE

In the end, we choose Option 1. We trade our former prime downtown location for extra space, a shorter commute for Sean, and a private garage. Now the nail-biting anxiety sets in as we try to lock in our mortgage rate. The rates change literally every day. Two days after our offer is accepted, we get a nasty surprise: There is a new 1 percent hike in mortgage interest rates for condominium units. So that 5.25 percent interest rate we used to calculate our estimated monthly payments will be 6.25 percent instead.

The earliest recorded mortgages date back to 1190 in England, when landowners would sell their land for a set fee, with no interest. Whatever the land produced would enable the buyer to pay the seller. *Mort* comes from the Latin word for "death," while *gage* means a pledge to forfeit an asset for nonpayment of a debt. The modern concept is not much different: We want to buy a house, but we don't have enough cash in hand to pay the full price, so we put down the cash we have and borrow the rest, using the house as collateral. There is a monthly payment, determined by the interest rate (usually fixed) and the lifetime of the loan (typically thirty years). At the end of that time, we will have paid off the principal loan plus the accumulated interest.

It's instructive to crunch the numbers and see firsthand why a mere 1 percent hike in the interest rate makes a significant difference on one's monthly payments. Let's round down the respective rates to 5 percent and 6 percent to make our calculations easier. If we took out a modest $100,000 mortgage at 5 percent, our payment would be $536.82 per month, compared to a $599.55 monthly payment at 6 percent interest. This assumes the interest is charged yearly. According to Mark Chu-Carroll, a computer scientist who blogs at Good Math, Bad Math, even this trivial difference can result in a higher monthly payment. We would only pay $525 per month at the 5 percent rate if the interest were calculated monthly. Just an extra $62 per month over 30 years adds up to roughly $22,320 in additional interest.

Fortunately, our story has a happy ending: We are able to negotiate our original estimated interest rate with one lender. It helped that we had a sufficient down payment. Early in the 1900s, aspiring homeowners in the United States were required to have a 50 percent down payment on a five-year mortgage. Because very few people could meet those conditions, fewer than 40 percent of the population owned their own homes, compared to nearly 70 percent today, when 20 percent down is more common for a thirty-year mortgage. How long would it take to save 20 percent of a $300,000 home? That is $60,000—not an easy sum to accrue on a standard living wage. But it is a simple matter to figure out how much we'd need to save each month to reach that goal within five years. We simply divide $60,000 by five to get our answer: $12,000 a year, or $1,000 per month. No calculus required.

However, that would only be the case if I took that money and stuffed it under a mattress. Common sense would dictate I deposit the funds in an interest-bearing account. Let's be optimistic and assume that account yields 5 percent interest. How does that change the time needed to save a down payment? I am depositing $1,000 per month, but the money that's been in the account for two years will have earned more money than the money I've just deposited.* So I have to add the $1,000 I deposit that first month and calculate how much interest will accrue in five years, and then do the same for the second monthly deposit, and the third, and the fourth, and so on.

Now it is a matter of adding together lots of smaller sums—or of taking an integral. Even though I am making monthly deposits, from a calculus standpoint, the funds are accumulating every instant. So for every interval of Δt (for

* This works in reverse on mortgage interest. Not all of your monthly payment goes toward paying off your principal. Most of it goes toward interest in the early years, because interest is always paid on the outstanding balance of the loan, which decreases over time as you pay down the principal.

time), we have accumulated $12,000 × Δt dollars, which stays in the bank for however much of that five-year period is left (5 years – *t*) and earns 5 percent interest. At the end of that five-year period, I will have saved $60,000 plus a bit extra in accrued interest—which I hope will be sufficient to cover the potentially exorbitant closing costs.

What if you don't have the required 20 percent deposit on a home—a common problem for those living in areas with especially high housing costs? Traditionally, you would be out of luck; no bank would approve your mortgage. There is very good reason for this. That down payment gives you equity in the house, the difference between your home's assessed value and the amount of money you still owe the bank. But then alternative types of mortgage loans became increasingly available, some allowing borrowers to take out a mortgage with as little as 5 percent down. The trade-off for the lower down payment is usually higher interest rates and thus higher monthly payments.

Then someone had the brilliant notion of offering adjustable rate mortgages (ARMs), in which the interest rate fluctuates over time, resetting to a new (higher) rate every few years. We have already seen that a small increase in the interest rate on a mortgage can make a huge difference in the monthly payment. The impact is even more dramatic with an ARM. Say you took out a loan of $100,000 at an adjustable rate over thirty years. You could easily afford the monthly payments at the introductory "teaser rate," which could be as low as 1.2 percent for the first two to five years: roughly $331 per month. But then the interest rate would reset and jump to 7 percent, and suddenly you would be paying $617 a month. Unless you had a corresponding increase in income, you would quickly fall behind in your payments. Worse, some of those ARMs were interest-only loans, for which people would pay just the interest and the principal never decreased.

Millions of people took on these risky loans; given the above, it's fair to ask, what the hell were they thinking? Chances are, they weren't doing the math. Or perhaps they believed that the value of their houses, and hence their equity, would continue to skyrocket, and they could sell their homes at a tidy profit before the interest rates reset.

But nothing can expand forever—except, perhaps, the universe—and those homeowners were gambling that they could get out before the market softened or collapsed outright. Several economists warned that the bubble would burst, but their dire predictions did little to dampen the enthusiasm at the height of the housing frenzy. All the classic bubble conditions were present: high demand, limited supply, and an influx of ready cash as banks relaxed their lending standards and made millions of subprime loans to borrowers who—in retrospect—should never have received loan approval because they couldn't afford the payments once the interest rates reset. When those buyers began to default en masse, the result was a record number of foreclosures.

VIRTUAL WEALTH

The fallout from Holland's tulip mania crash was limited to a few overly enthusiastic traders and wealthy collectors. That's because the Amsterdam Stock Exchange back in 1630 had the good sense not to get involved with the rampant speculation in tulip bulbs, marginalizing the economic impact when the bubble burst. Most Dutch traders were able to negotiate settlements for their debts, although the price of bulbs continued to fall for decades after the crash. Financial ruin hit those who had invested elsewhere while relying on the profit they expected to make on their tulip bulbs to pay those debts—profit that never transpired.

That was the problem with the housing bubble: People speculated on the market, tapping into the equity on their homes to finance other projects—a new car, a lavish vacation, a kitchen remodel, an investment in a second rental property, or a vacation home. When the market crashed and their home values plummeted, those home owners found themselves owing more to the banks than their homes were worth. They had *negative* equity. Furthermore, investment banks had packaged those mortgages into complicated financial instruments that were sold to investors around the world, so when the waves of foreclosure hit, the massive losses incurred over a short period of time brought the global economy to its knees.

Economists are going to be analyzing this housing market crash for decades before they fully understand how and why it happened. But anyone observing the virtual economy in the online game Second Life could glean some valuable insights, according to Cornell University economist Robert Bloomfield. He believes virtual economies like those in Second Life can provide useful simulations of the patterns of free markets—and the consequences of failing to self-regulate. In Second Life, players can buy virtual currency with their real-world dollars—250 "Linden dollars" roughly corresponds to one U.S. dollar. They buy and sell goods and services and engage in online investment schemes without all the pesky regulations hampering the free market in "meat space."

And therein lay the problem. In 2007, an in-game virtual investment bank, Ginko Financial, collapsed. The bank promised investors a whopping 40 percent return on their Linden money and made loans to other players at equally exorbitant rates. When those players failed to repay their virtual loans, investors panicked and made a run on Ginko to withdraw their funds, quickly outstripping the bank's reserves. Nor were the losses purely virtual, since Linden dollars were purchased with real currency: Investors collectively lost the equivalent of 750,000 U.S. real-world dollars.

Second Life creator Linden Lab responded by banning any virtual banks promising interest rate returns on deposits to investors. One year later, in the wake of the mortgage meltdown, revered financial titan Alan Greenspan reluctantly came to a similar real-world conclusion: Lending institutions cannot be trusted to regulate themselves—not because the free market doesn't work, but because certain unscrupulous people cheated and "gamed" the system. It is human nature that is at fault, more than free-market economics. It makes a strong case for factoring irrational human behavior into any viable economic model. In fact, the burgeoning new field of behavioral economics focuses on studying how and why human beings don't always act in their best self-interest.

The parallels to our real-world economy are admittedly imperfect, but the economic lessons drawn from Second Life are compelling, because it is a model built from actual human behavior—raw data—not a programmed computer simulation. People do not always behave rationally (or nobly), and many economic theories fail to take this into account. Jonah

Lehrer, author of *How We Decide*, asserts that the problem lies less with the actual models and more with the human brain. "People love models, especially when they're big, complex, and quantitative. Models make us feel safe," he writes. "They take the uncertainty of the future and break it down into neat, bite-sized equations. But we become so focused on the predictions of the model that we stop questioning the basic assumptions of the model. Instead, confirmation bias seeps in and we devote way too much mental energy to proving the model true."

Models can still yield intriguing insights. Reginald Smith, an analyst with the Bouchet-Franklin Research Institute in Rochester, New York, decided to map the spread of the collapse from its start in the housing markets of California and Florida in 2007 through October 2008. He found that the problems first emerged in housing stocks, then spread to finance stocks and mainstream banks before hitting the broader stock market in general. While his analysis didn't shed much light on the why of the collapse, he noticed that his data bore a strong resemblance to a different kind of model: that used by scientists to chart the spread of forest fires, fashion trends, . . . and disease. Mathematically speaking, the credit crisis looks like an epidemic, wiping out wealth the way the Black Death decimated the population of medieval Western Europe.

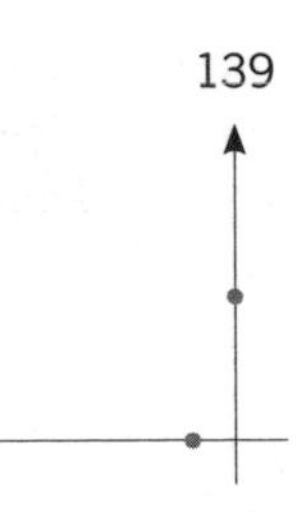

6

A Pox upon It

> Population, when unchecked, increases in a geometrical ratio. Subsistence increases only in an arithmetical ratio.
>
> —THOMAS ROBERT MALTHUS

It is a truth universally acknowledged that a young man in possession of good fortune must be in want of a well-stocked arsenal—at least if you're one of the eligible young men who populate the blood-soaked satire *Pride and Prejudice and Zombies*. Author Seth Grahame-Smith invented an alternate history for Jane Austen's much-beloved novel, in which a mysterious plague sweeps through the peaceful village of Meryton, turning residents into the walking dead, famished for fresh brains upon which to feast. Under those circumstances, any woman who can wield a weapon as well as a witty bon mot is doubly attractive.

In bizarro Meryton, Elizabeth Bennett and her four sisters make up an elite zombie-fighting unit, well versed in the usual feminine accomplishments: music, needlepoint, watercolors, and of course, martial arts and weapons training. Their mission: wiping out the undead menace while finding suitable wealthy husbands. The very first ball at Netherfield is overrun by "unmentionables" who feast with abandon on the hapless guests, "sending a shower of dark blood spouting as high as the chandeliers." Female characters debate whether or not it is "unladylike" to carry a musket (Elizabeth favors a *katana*, or samurai sword), and couriers routinely get eaten by zombies while relaying messages between houses. The local militia comes to town to exhume and destroy dead bodies, hoping to control the outbreak. And Elizabeth must defeat Lady Catherine de Bourgh and her merry band of ninjas to win the right to marry Darcy.

Grahame-Smith felt Austen's original text was a natural fit for zombie horror. "You have this fiercely independent heroine, you have this dashing heroic gentleman, you have a militia camped out for seemingly no reason whatsoever nearby, and people are always walking here and there and taking carriage rides here and there," he told the Daily Beast. "It was just ripe for gore and senseless violence." And ninjas—don't forget the ninjas. It makes even more sense when one considers that Regency England was no stranger to deadly outbreaks of disease and that the modern zombie genre pioneered by George Romero's *Night of the Living Dead* routinely treats the spread of rampant zombification as an epidemic. As such, zombies provide an excellent case study in epidemiology.

Epidemiologists study the rate at which disease outbreaks spread and how various intervention strategies—vaccination or quarantines, for example—can help slow the transmission rate. They study this in the context of general population dynamics: the number of infected individuals and the rate at which a population grows or declines are connected. If there's a varying rate of change between two connected factors, there must be a derivative to be taken somewhere. So calculus is very useful in epidemiology and therefore in the analysis of zombie outbreaks. It just so happens that nature has its own microcosmic version of a zombie epidemic, which lends itself very nicely to illustrating a fundamental epidemiological model.

A FUNGUS AMONG US

Deep in the forests of West Central Africa lurks a species of parasitic fungus that targets a particular kind of ant. The fungus belongs to the *Cordyceps* family, scattering spores into the air, which then attach to the ant's body to germinate. The spores work their way inside the poor insect's body, sprouting long tendrils called mycelia that eventually reach into the ant's brain and release chemicals that make the ant the fungus's zombie slave.

The chemicals change how the ant perceives critical pheromones, altering its behavior. In this case, the ant feels less inclined to devour delicious brains and is instead compelled to climb to the top of the nearest plant and clamp its tiny jaws around a leafy stem. It is the fungus that plays the role of zombie now, devouring what little remains of the insect's brain, then sprouting through the ant's head as one final indignity. Those sprouts burst and release even more spores into the air, which go forth to infect even more unsuspecting ants. The entire horrific process can take four to fourteen days. Fear the fungus, my friends.

There are over four hundred different species of *Cordyceps* fungi, each targeting a particular species of insect, whether it be ants, dragonflies, cockroaches, aphids, or beetles. Consider *Cordyceps* an example of Nature's own population control mechanism to ensure that ecobalance is maintained. The fungus proliferates when there is a large supply of hosts—that is, when the ant population flourishes and becomes so large that it threatens to overwhelm the resources available to the colony. As more ants fall victim to zombifying spores, their numbers dwindle until (a) there are once again sufficient resources to support what remains of the colony, and (b) there are far fewer ants available to serve as hosts, making it more difficult for the fungi to reproduce, so their numbers dwindle as well. And the whole population growth-and-decline cycle begins all over again. That is the essence of population dynamics in a nutshell.

An English clergyman named Thomas Robert Malthus was one of the earliest pioneers in modeling population dynamics. Malthus was born with a harelip and cleft palate—defects that ran in the family—and was intensely self-conscious about his appearance as a result. He had an unremarkable childhood in the Surrey countryside, earning a mathematics degree from Cambridge University before being ordained as an Anglican curate.

Malthus bemoaned the decline of living conditions in late eighteenth-century England and observed that in nature, plants and animals were capable of reproducing at far greater rates than the surrounding resources could support. This led him to develop his classic theory on population: If human population were allowed to grow unchecked, it would do so exponentially, and we would all too quickly outstrip our limited resources for subsistence. He believed this fundamental truth had been obscured by catastrophic events like disease, famines, or wars, which serve periodically to cull the herd, so to speak. "Epidemics, pestilence and plague advance in terrific array, and sweep off their thousands and ten thousands," he wrote with considerable dramatic flourish. "Should success be still incomplete, gigantic famine stalks in the rear, and with one mighty blow, levels the population with the food of the world."

In 1798, Malthus published *The Principle of Population*, in which he outlined his model for population growth. It's based on the notion that the population for a given generation is dependent on the size of the previous generation, and that this number will be a multiple. We can denote population size (p) as a function of time (t), where t can represent any unit of time we choose: days, months, years, and so forth. The key parameter is known as the Malthusian factor (r), denoting the multiple that determines the growth rate. We can plot different values for p when $r = 1.19$, 1.20, and 1.21 to see how a slight change in the value for r (denoted by the variable a in the figure) results in significant differences in the overall population size. The resulting graph produces three different exponential growth curves. Even a difference as small as 0.02 causes population to double after 40 units of time (whether it be 40 days or 40 years).

Maybe it had something to do with being one of eight children, but Malthus's proposed solution to overpopulation included restricting the family size of the lower classes to ensure that parents did not produce more children than they could support. It sounds more elitist than he perhaps intended: Malthus thought having too many children doomed the lower

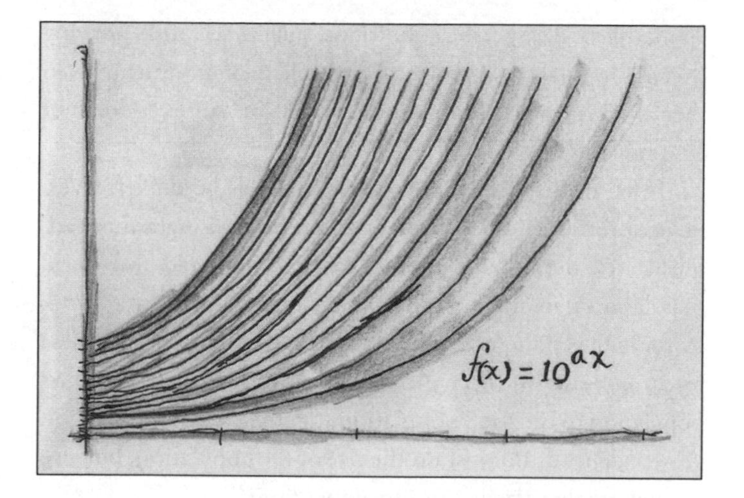

classes to poverty, making it impossible for them to rise above those conditions and improve their lot in life. Then again, he also flirted with the notion of eugenics—a term not coined until 1883—by pondering whether the techniques of animal husbandry might be applied to breed out undesirable qualities in people, although he didn't feel this was a realistic goal: "As the human race, however, could not bc improved in this way without condemning all the bad specimens to celibacy, it is not probable that an attention to breed should ever become general." (Yes, even in the 1800s, people realized that abstinence alone is not a viable solution for family planning.)

This sort of pessimistic thinking did not win Malthus any popularity contests at a time when fervent social reformers preached the gospel of erasing all the ills of man if only one could implement the proper social structures. The model is not without merit, but the Malthusian growth equation is only applicable under specific conditions, such as scientists growing bacteria in a lab in a perfectly controlled environment. Even then, while growth occurs exponentially for a time, it does not continue forever.

It fell to his contemporary, the Brussels-born Pierre Verhulst, to improve upon the basic idea by devising a more sophisticated model that more accurately reflected real-world population dynamics. Verhulst said there are forces at work to prevent exponential growth in the population, and these forces increase in direct proportion to the ratio of the excess population to the total population. In other words, population growth depends not just on the size of the population, but also on how far that size is from its upper limit.

The crux is something called carrying capacity (which we can denote by K): the maximum population size that any given habitat can support. If the population of ants starts to double every year—grows exponentially—there will be two thousand ants the first year, and the next year there will twice as many. But there is a limited supply of food and other resources, so if that exponential growth rate continues unchecked, the population of ants will rapidly consume all the available resources. Exponential growth simply cannot be sustained indefinitely. Once the food runs out, the ants will begin to die out too. Verhulst's model* shows that if the population is less than the maximum, the population will increase rather steeply because people have lots of food. But then when it gets closer to the maximum sustainable population, the rate at which the population increases slows down.

* See appendix 2, "Calculus of the Living Dead," for a detailed breakdown of this type of calculus problem.

Look closely and you will recognize the telltale sign of a derivative. In the Verhulst model, the derivative of the population with respect to time—that is, the rate of change in population (p) or the number of additional people over time—would be proportional to the number of people at time 0 (now) multiplied by the maximum sustainable population minus the current population. If the population actually followed that equation, it would start out low and show exponential growth at first. But then the rate of growth would begin to slow as it approached the maximal population, eventually leveling off to become stable as it reaches the carrying capacity for that particular habitat. Plot this out on a graph and you end up with a smooth S-shaped, or sigmoid, curve.

Say you're part of a colony of a particular species of ant, going about your business in the forest: gathering food, doing your little communication dance, and of course, reproducing. We can use the derivative to analyze the rate at which your little colony is adding to its population. We'll keep things sim-

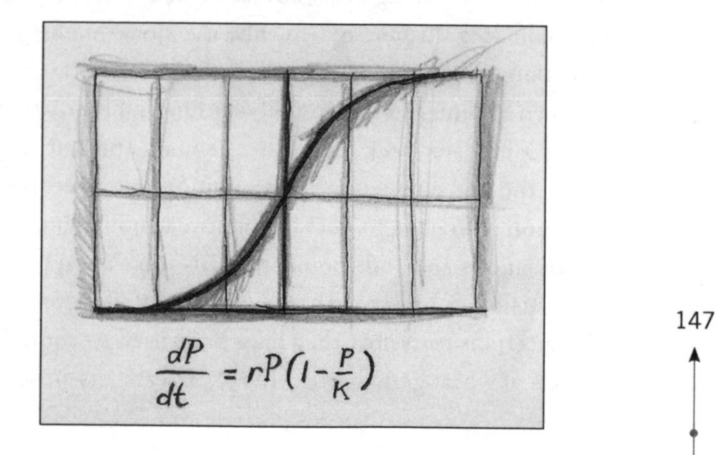

ple by assuming an initial population of 100, which increases to 120 ants after one year. How long will it take your little ant colony to grow from 100 to the critical threshold, or carrying capacity (K), of 300 ants? Just plug in the relevant numbers to the Verhulst equation: 100 for the initial population, and (in this case) a growth rate per year (r) of 1.2. The answer: six years.

The Verhulst model is useful for limited applications, but the realities of population dynamics are far more complex, with innumerable variables. Even the carrying capacity (K) is not a constant (fixed over time); it fluctuates depending on conditions. Furthermore, instead of being continuous, as in the Verhulst model, population change often occurs in discrete shifts. Instead of the population changing continuously in tiny increments each day, there may be a major event that will cause the population to either explode or rapidly decline. An earthquake that wipes out an entire village would result in a sudden rapid decrease, while an influx of immigrants or refugees would give rise to a sudden spike in population. Then we are no longer dealing with a straightforward calculus problem, but something akin to a chaotic system, like the stock market's wild fluctuations, making predictions extremely difficult.

When it comes to our zombifying fungi, the situation resembles a predator-prey model: as the fungi (predators) proliferate, the ant population (prey) diminishes; when the ant population flourishes, so does the predator population, so you have equations for both populations. Nature always finds a way to maintain balance. These fungi are so effective at controlling certain pests that they have been used to control the numbers of wheat grain beetles. In fact, researchers are investigating the use of one particular species of fungus (*Metarhizium anisopliae*) against African mosquitoes to control the spread of malaria, because the disease is often spread through mosquito bites. That's another useful application of calculus: assessing the rate of the spread of a disease, and determining how effective various intervention strategies might be.

MATH IN THE TIME OF CHOLERA

Cholera is a nasty way to die. It starts with horrible bouts of vomiting and diarrhea and a slowed pulse, plus cramps. Those cramps become more severe as the disease progresses, the victim's entire body convulsing in pain. Eventually the lips, face, hands, and feet turn blue, purple, or even blackish in hue. The skin becomes cold and damp. Respiration slows, but instead of a telltale death rattle in the throat, victims often die quietly, with a whimper. At least the disease progression is rapid, so one's misery is short-lived. That's about all that can be said for it.

In the nineteenth century, England's physicians, scientists, and political leaders watched with trepidation as *cholera morbus* moved from India through Eastern Europe to Germany and the shores of England, officially "arriving" in London in 1831. Cholera killed over 10,000 people in one year alone. In 1854, London's Soho District was hit by an especially virulent outbreak of the disease, killing 127 people in the first three days. By the time it was over, 616 people had died.

The means by which a disease spreads throughout a population is known as a vector; the most common vector is person-to-person transmission, such as with the flu or measles—or a zombie bite. With cholera in the nineteenth century, the vector was less clear. Medical opinion was divided, because the evidence was contradictory, sometimes indicating transmission through contact, sometimes indicating transmission through squalid unsanitary conditions. The streets of Soho in the 1850s were filled with animal droppings, runoff from slaughterhouses, and primitive sewers. Had anyone checked under the floorboards of their cellars, they would have found fetid cesspits.

The man who solved the mystery was Dr. John Snow, a pioneer of modern epidemiology. He lived locally, on Fifth Street, and monitored the epidemic's progress on-site. He was convinced that cholera was spread by a poison passed from victim to victim through tainted water; he'd already traced an earlier outbreak of contaminated water supplied by the Vauxhall Water Company. But authorities didn't believe him, and the water company refused to admit culpability. He figured this was his chance to prove his theory was right.

Snow patrolled the district, interviewing the families of those who had died, and found that nearly all the deaths had occurred near a water pump on the corner of Broad Street and Cambridge Street—the epicenter of the outbreak. Houses closer to an alternate pump had only experienced ten deaths, and five of those were schoolchildren who occasionally drank from the Broad Street pump. Ever the scientist, Snow took a sample of the pump's water, examined it under a microscope, and noted that it contained "white flocculent particles," which he deemed the cause of the infection.

The Board of Guardians in St. James Parish reluctantly followed his advice and removed the pump handle as an experiment. The spread of the disease stopped dramatically. There were still a few unexplained deaths from cholera that appeared unrelated to the Broad Street pump. The most damning was a widow who lived in Hampstead, and her niece, neither of whom lived anywhere near Broad Street. Snow proved quite the detective: He found that the widow had once lived in Broad Street and liked the taste of that well water sufficiently that she had a servant bring her back a large bottle from it every day. The last bottle had been fetched on the day the Soho outbreak began.

Yet authorities were still doubtful of Snow's findings. A local vicar, Reverend Henry Whitehead, thought the outbreak was the result of divine intervention—a very vicarlike approach to human calamity—and set about "proving" his case. In the end, Whitehead actually helped confirm a single probable cause of the outbreak: A young child living on Broad Street had been ill with cholera symptoms, and the child's soiled diapers had been soaked in a tub of water that was then emptied into a cesspool three feet from the Broad Street pump. Underground leakage did the rest.

How do we model an outbreak of a disease? Let's assume that a nasty flu virus strikes a university dormitory. The rate of infection will vary, depending on the nature of the disease and how it is transmitted. The flu is spread when an infected person, during the contagious period, coughs or sneezes near another person or touches another person. We can chart how the number of infected people (I) changes over time (t)—in other words, I is a function of t, and for our purposes t will be measured in days. For epidemiology, there are two other parameters: how many people an infected person can infect per day, or rate of infection (r), and the rate at which the outbreak fizzles, as infected people recover—or die (a). So there will be a single equation for $I(t)$, in which r and a will appear as parameters.

The end result is almost always the same: As more people recover or succumb and as precautionary measures kick in—quarantine, hand-washing, or just removing the handle of the offending pump—there are fewer new cases of infection. When r is less than 1, each infected person is, on average, transmitting the virus to fewer than one other person. This will not be sufficient to sustain the outbreak, and it will end. As for the flu, so for cholera.

Fourteen years after Snow's discovery, a cholera epidemic hit Buenos Aires, Argentina. An account of the outbreak can be found in Charles Darbyshire's *My Life in the Argentine Republic 1852–1894*. He moved his household to the countryside because he worried about the unsanitary conditions of town life, having seen the impact of cholera in London before he came to Argentina. He described the conditions in alarming detail:

> I felt positive that sooner or later there must be an epidemic. There was no drainage. The soil on which the houses were built was becoming infected. The defecations, the waste water from kitchens, etc. went into wells 30 feet deep in the back patios. When one of these wells became full of filth and could hold no more, what was called a *sangria* (a bleeding) was made. A well was sunk to the same depth . . . and the sangria took place by pushing an iron bar through the full well . . . as the old well began to drain into the new.

This went on for years, and some of the patios in the old houses were honeycombed by wells.

Darbyshire's fears proved well founded when an epidemic broke out in the summer of 1868, brought about (he believed) by Brazilian ships tossing the bodies of those who had died from cholera into the River Paraná, contaminating the water supply. People fled to the countryside, bringing the disease with them, and Darbyshire advised his neighbors not to drink the water unless it was boiled, to bury all refuse, and to keep floors and patios clean. His own household did not contract the disease, which lent credence to his advice. Despite all the deaths, there was one positive outcome: The Argentine government overhauled the city's drainage system and installed a proper water supply.

Darbyshire correctly identified a contaminated water source as the source of the outbreak. Initially the disease spread at a very rapid rate, and people panicked. With no quarantine in effect, those already infected brought the contamination to the countryside. Had that first city been quarantined, the outbreak would have been contained as the rate of removal (*a*) increased. Just as with an economic market, at some point, a critical threshold is reached, and the exponential growth rate of removal would level off as fewer and fewer people remained to be infected. Darbyshire had the foresight to put protective measures in place that limited the spread of infection. That caused the rate of removal to increase even faster, and thus the outbreak leveled off and died out much more quickly.

In the case of cholera, there was a single vector: the Broad Street pump, or, more specifically, the white flocculent particles contained in the water that came out of that pump. The disease spread to whomever drank from that particular source. A disease like the Black Death is much more complicated to model because there is more than one vector.

MASQUE OF THE BLACK DEATH

One week before Christmas in 1664, a comet streaked across the sky over England. Astrologers claimed it was an omen of impending apocalypse. One William Lilly predicted that this, combined with a lunar eclipse in January 1665, would bring "the sword, famine, pestilence, and mortality or plague." Lilly was really hedging his bets—why not throw in a prediction of a zombie invasion or an asteroid strike while he was at it?—but his dire prediction came partially true. Pestilence was common in the rat-infested urban centers of England, and this led to a deadly outbreak of bubonic plague in London in the summer of 1665.

By October, one in ten Londoners had succumbed to the disease—over sixty thousand people. The government banned public meetings, but the epidemic spread to Cambridge, where the young Isaac Newton was in his second year of studies. The university closed, and Newton was forced to return to his country home in Grantham for over a year until the plague had run its course and the university opened its doors again in April 1667. And in that short time, he invented calculus, with no idea that it would one day be applied to study the spread of disease.

This was not the plague's first appearance. Back in the Middle Ages, the plague decimated Western Europe, wiping out

roughly one third of the population, some 25 million people. It could sweep through a region and wipe out entire villages in a matter of weeks. During the 1630s, various outbreaks of plague killed half the populations of affected cities. Similar numbers perished in an outbreak in Holland in the 1660s: A thousand people were dying each week in Amsterdam at the height of the outbreak. And the plague significantly culled the population of France during an outbreak between 1647 and 1649.

The plague spread rapidly and was so virulently infectious that even doctors feared treating victims. At the time, they believed the disease spread via "bad air," or miasmas. Those who did treat patients took what precautions they could, donning large beaked hats made of bronze and stuffing the "beak" with strong herbs and spices to purify the air the doctor breathed. (As an added bonus, the aroma from the herbs helped mask the stench of rot that invariably accompanied the plague.) Plague doctors dressed in pants and a long gown and wore leather gloves, as well as crystal eyepieces for added protection—

anything to ward off contamination, even though the source of the plague was not identified until the nineteenth century. All clothing, even undergarments, were doused in camphor oil or treated with wax to further seal the doctor from bad air.

These precautions might have been partially effective. We now know that plague is caused by a bacillus called *Yersinia pestis* and is spread by rodents and their fleas to humans.* Protecting the eyes, nose, and mouth made it harder for *Y. pestis* to get into the body via mucous membranes, and coating one's clothing with wax made it more difficult for fleas to penetrate to the skin and transmit the disease with their bites. And the herbs stuffed into the beak of the mask at least partially blocked breathing holes, so the doctor would be less likely to inhale the bacillus. Their Achilles' heel was actually the ankles, which remained exposed and therefore vulnerable to flea bites.

The credit for this momentous discovery goes to a French scientist named Alexandre Yersin, a former student of Louis Pasteur.† Yersin went to Hong Kong in 1894 to investigate an outbreak of the plague there. He extracted some of the pus from a dead soldier's bubo (swollen lymph node) and injected it into guinea pigs; all the guinea pigs died. He examined the pus from both the dead soldier and the doomed rodents and noticed both samples contained the same type of bacte-

---

* An alternate theory proposes that while *Y. pestis* is responsible for modern outbreaks of plague—and yes, there are still outbreaks around the world, mostly concentrated in Africa—the Black Death that ravaged Western Europe in the fourteenth century was caused by something like anthrax or an Ebola-like virus. The evidence is sketchy, however. An analysis of the remains of early plague victims in France showed DNA from *Y. pestis* and none from anthrax, for example.

† It was discovered simultaneously by a Japanese scientist named Shibasaburo Kitasato, but the microbe is named after Yersin.

ria. Yersin also noted the large number of dead rats around the city, examined those bodies, and once again observed the same bacteria. Conclusion: *Y. pestis* was the culprit for the spread of plague.

Yersin did not determine the means of transmission, however. That honor fell to his fellow scientist, Paul-Louis Simond, who experimented with infected rats and fleas. He noticed that even if he placed an infected rat into a jar with healthy rats, the healthy ones only became sick if fleas were present. Just how virulent is this plague-causing *Y. pestis*? In lab experiments, mice died after being infected with just 3 bacilli; your average flea can transmit 24,000 in a single bite.

Plague has many different vectors: It can spread person to person or via the rats and fleas; which type you get depends on how the bacillus invades your body. The Black Death came in three forms: bubonic, pneumonic, and septicemic. After the bite of an infected flea, the first site of infection is generally the lymph nodes. In this form, bubonic plague, your lymph nodes swell to form enormous buboes. Lancing the buboes releases oozing, foul-smelling pus. The bubonic plague was the most common form, with a mortality rate of 30 to 75 percent. In addition to enlarged and inflamed lymph nodes around the armpits, neck, and groin, victims were subject to headaches, nausea, aching joints, high fever, and vomiting, and symptoms took from one to seven days to appear.

The pneumonic plague, infecting the lungs, is particularly virulent, capable of killing an infected person within twenty-four hours. You would catch this merely by breathing *Y. pestis* into your lungs. The mortality rate for the pneumonic plague was 90 to 95 percent (if treated today, that would be reduced to

5 to 10 percent). Symptoms included slimy sputum (a saliva-and-mucus concoction) tinted with blood. As the disease progressed over one to seven days, the sputum turned bright red.

If *Y. pestis* entered your bloodstream directly through the bite of a flea or via a cut or sore in contact with diseased tissue, you would get septicemic plague and would be almost certain to die. The septicemic plague was the rarest form of all, but the mortality rate was close to 100 percent; even today there is no treatment. Victims ran a high fever, and the skin turned deep shades of purple, almost black, hence the name Black Death. Victims usually died the same day symptoms appeared; in some cities, as many as eight hundred people died every day.

Septicemic plague was rarer than the other two forms of plague because people died so quickly that they had little opportunity to transmit the disease to others. That's why any good epidemiological model must take into account the latency period between infection and death. Pneumonic plague was easily transmitted from person to person, but death usually occurred within a day or two, so it, too, did not propagate as rapidly. Bubonic plague gets it just right, from the perspective of *Y. pestis*, whose sole purpose is to infect as many hosts as possible. It is not as virulent as pneumonic plague. Once infected, the victim could appear healthy for as long as a week, merrily passing the disease on to others, and death occurred much more slowly.

"Because of its infectious nature, the disease may be spread by apparently healthy people who harbour the disease but have not yet exhibited the symptoms," Daniel Defoe wrote in *A Journal of the Plague Year*, which appeared in 1722. "Such a person was in fact a poisoner, a walking destroyer perhaps for a week or a fortnight before his death, who might have ruined those that he would have hazarded his life to save." Defoe may have been writing about the real-life plague that decimated London in the 1600s, but he could just as easily have been describing Grahame-Smith's alternate version of the village of Meryton, where residents who were bitten would seem normal but were in fact gradually turning into zombies.

ASSUME A SPHERICAL ZOMBIE

Pride and Prejudice and Zombies is rife with graphic battle scenes, as Elizabeth Bennett travels the countryside with her aunt and uncle, leaving a path of zombie casualties in her wake. She teams up with Darcy to defeat a horde of zombies at his Pemberley estate, and after accepting his proposal of marriage, the newly engaged couple dispatches one final group of zombies to plight their troth. But is all this bloody violence toward zombies really necessary? Can't humans and zombies learn to get along and coexist in harmony?

According to a 2009 paper by a group of Canadian epidemiologists: no way, nohow. The lead researcher is Robert Smith?* of the University of Ottawa, who specializes in modeling the spread of infectious disease. He and three students adapted their models to the spread of a fictitious zombie infection, starting out with a simple model and gradually adding elements to make it more complex.

* Yes, there really is a question mark at the end of his name. He changed it to distinguish himself from the zillions of other Robert Smiths in the world, including the lead singer of the Cure: "It's been twenty years now and sadly his career shows no sign of drying up," the epidemiologist laments.

"The key difference between the models presented here and other models of infectious disease is that the dead can come back to life," the authors write, tongues firmly in cheeks. According to Smith? and his students, people fall into three basic categories: susceptibles (S), those who are not infected; zombies (Z); and removed (R), susceptibles who have died of other causes. The key factor is not the actual numbers in each category, but how those numbers change with time as new zombies are made and existing zombies are killed. Anytime we have a rate of change, we have a derivative situation on our hands. The rate of change in zombies is the net increase or decrease in their numbers during a given period of time.

There are well-established rules governing the zombification process.* Zombies can be killed by cutting off their heads and destroying their brains. Susceptibles can become zombies if they are bitten by one, but zombies can also be created by resurrecting the removed—those who are already dead. If we have six humans turned into zombies every hour and four dead people resurrected into zombies every hour, the result is ten new zombies every hour. Now let's say we manage to kill three zombies every hour. The net result is an increase in the zombie population of seven zombies per hour. And at that rate, there is no chance of maintaining what's known as an endemic state—one of peaceful coexistence, or at least a comfortable equilibrium.

Smith?'s model doesn't end there; that's just the process for

* The researchers' model is based on the zombies as featured in *Night of the Living Dead*, as opposed to the modern take depicted in, say, *28 Days Later*, wherein the zombies were smarter and moved faster than the lurching, drooling classic monsters whose sole purpose is to devour delicious brains.

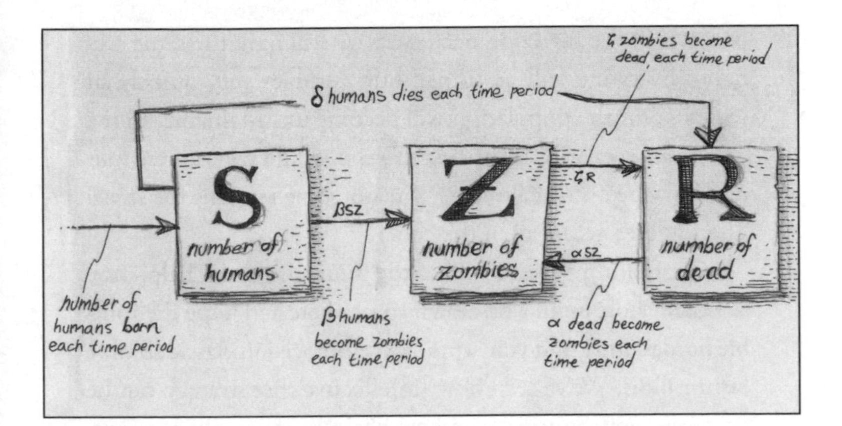

calculating the rate of change in the zombie population. We also have to run equations for how the number of dead and the number of uninfected humans change, which means factoring in the birth and death rates of humans as well. This is called a coupled system of ordinary differential equations, which is really just a fancy way of saying that the system must be described by not one, but three connected equations: one for how the number of humans changes, one for how the number of zombies changes, and one for how the number of dead changes. Furthermore, Elizabeth Bennett's good friend Charlotte has been bitten but is not yet a zombie, although doomed to become one in a matter of weeks—as good an explanation as any for her marriage to the odious Mr. Collins. Smith? and company call these people Latents, giving us a total of four coupled equations. The coupling occurs because the same variables appear in all four equations—or, practically speaking, because the different populations interact with one another.

Assuming the zombie infection occurs quickly, the birth and death rates of humans will be insignificant during the time over which the infection occurs, so we still have the same scenario: Everyone will be turned into zombies very quickly, at which point the population will become unsustainable. In the worst-case scenario, Smith? estimates it would take a mere four days to wipe out the humans. The outcome remains the same: The zombies get us all in the end.

Quarantining the few healthy humans could help—the standard "hole up in a basement somewhere and hope the zombie hordes don't find you" approach employed in classic zombie horror films. We've seen how (in)effective that strategy can be onscreen, and Smith?'s numbers back up those observations. However, another study by an Italian scientist named Davide Cassi implies that hiding out at the mall (à la *Dawn of the Dead)* could vastly improve one's chances of survival. Cassi wasn't analyzing zombies specifically, but his version of a predator/prey model applies to any kind of "predatory random walker": organisms (like zombies!) that stumble around without any obvious purpose or direction, destroying any human that comes into their path. The larger and more complex the structure—such as a large mall with many twists and turns—the lower the chances that the predator will stumble upon the prey.

Alternatively, we can quarantine the zombies by herding them into some sort of holding pen, but if we don't isolate enough of them fast enough, once again, the zombies will win. Both options are rather passive strategies, and most likely will only postpone the inevitable annihilation of the human race.

Smith? and his students suggest that our only hope is an "impulsive eradication" scheme. A series of fierce, concentrated attacks could sufficiently cull the number of zombies over time so that the outbreak would finally die out. "The most effective way to contain the rise of the undead is to hit hard and hit often," the paper concludes. "As seen in the movies, it is imperative that zombies are dealt with quickly, or else we are all in a great deal of trouble."* Enter the Bennett sisters and respective paramours, with their wild, weapon-wielding ways, to make quick work of any rampaging zombie hordes.

Applying epidemiological modeling to a zombie invasion might seem silly, but it is not very different from modeling the spread of swine flu or the HIV virus. In November 2009, Smith? published another paper in the open-access journal *BMC Public Health*, arguing against spending $60 million in funding to combat the spread of HIV over fifteen to twenty years. Smith? recommended a far more aggressive five-year program—a variation on his "impulsive eradication" scheme for combating zombies—insisting that a gradual approach is doomed to fail because HIV/AIDS spreads so rapidly through travel and migration.

Smith?'s group also studies the kinds of slow-moving, chronic diseases in less developed countries that tend to be neglected by newspapers and funding agencies alike: things like leishmaniasis and dracunculiasis—both parasitic diseases that give rise to festering skin sores, among other symptoms—which can have long-term socioeconomic impacts on large populations. Dracunculiasis, or guinea worm disease, is particularly nasty. Drinking contaminated water will introduce the larva into your body, where it will hatch and grow for about a year until it forms a blister on your skin, which then ruptures so the

* The blog Southern Fried Science adapted Smith?'s models to a new scenario: pitting zombies versus vampires to determine which species would be most likely to survive. They concluded that zombies would eventually rule the earth in that scenario, unless the vampires and humans joined forces against the zombies.

worm's wriggling form sticks out.* With these types of diseases, as with zombies, the infected don't die: They live on, and thus have far more opportunity to transmit the disease to others.

SIX DEGREES OF ZOMBIFICATION

Most epidemiological models follow the basic format of separating the host population into those who are susceptible, infected, or immune to a particular pathogen. The assumption is that the rate at which new infections occur is proportional to the number of encounters between susceptible and infected individuals. That reproductive ratio doesn't merely depend on latent and infectious periods, but also on how much contact there is between those who are infected and those in the healthy-but-susceptible population.

This means that social networks play a big role in how quickly (or slowly) an outbreak propagates. The good citizens of Meryton go to balls, congregate in drawing rooms, and visit friends and relatives in other townships for a fortnight or more, providing ample opportunity for zombie infection to spread. So the more we know about the social networks involved in an outbreak, the better we can refine our epidemiological models.

Social networking is related to the small-world phenomenon, better known to most of us as "six degrees of separa-

* Think you can ease that painful burning sensation by soaking it in water? Bad idea. That just makes the adult female worm release hundreds of thousands of her larval spawn, further contaminating the water supply. The only way to get the worm out of your body is to wait until it pokes its head through the blistering skin, then wrap it around a stick and gradually pull it out. That process takes at least a month – a long, very uncomfortable month.

tion" and epitomized by the popular game Six Degrees of Kevin Bacon, in which players try to make a series of connections to the actor based on those who have been associated in some way with his movies. In the original 1967 study by psychologist Stanley Milgram, information packets were sent to randomly selected people in Omaha, Nebraska, and Wichita, Kansas, containing a letter describing the purpose of the experiment, providing basic information about the target contact in Boston, Massachusetts, and asking them to forward an enclosed letter. If the recipient knew the target, he or she would forward the letter directly. If not, the recipient would forward it to a friend or relative more likely to know the target. While the number of connections it took for the letters to reach the target varied, the average was around 5.5—hence, six degrees of separation.

Milgram's study fell into some disrepute when it was revealed that his famous experiment and conclusions were based on a minuscule data sample. In one experiment, out of sixty letters, fifty people responded to his challenge to forward the letter via their social networks, but only three letters eventually reached their destination. A far greater number of people didn't bother to participate in the experiment at all. That said, the study does offer intriguing evidence that smaller communities, such as those of actors and mathematicians* are densely connected by chains of personal or professional associations.

How do we even begin to track those interconnected chains? Nathan Eagle, an engineer at MIT's justly renowned

* "What's your Erdős number?" in honor of mathematician Paul Erdős, replaces "What's your sign?" as the pickup line of choice in math and science departments. Erdős is the Kevin Bacon of mathematics.

Media Lab, studies social networking phenomena using an unusual approach for data collection: cell phones. Mobile phones, with their GPS tracking components and call logs, make fantastic behavioral "sensors," providing a far more accurate record than asking people to record their own behavior in a diary—the traditional methodology for such studies. Self-reported data is notoriously error-prone, in part because people have faulty memories—or are not being entirely honest in their reportage.

In one study, Eagle and his colleagues provided cell phones to ninety-four test subjects—all students or faculty at MIT—loaded with special software that kept track of their location and logged all calls made and received among those phones. As a control, the study also included self-reportage, with subjects identifying which of the other subjects were friends, acquaintances, or strangers. Based on the calling patterns, the researchers were able to identify correctly whether two given subjects were friends or strangers more than 95 percent of the time.

That is just one study. Based on three basic parameters—a user's activity, location, and proximity to other users—Eagle says it is possible to accurately predict someone's future behavior based on limited observation of their current behavior. And because transmission of disease is strongly correlated to social networks and proximity to others, his method is extremely useful for epidemiological modeling.

According to Eagle, the typical epidemiological model rests on an erroneous assumption: that the probability of infection is equal for all, that is, the population is well mixed. But social networks are much more complex than that; there is noticeable clustering of social contacts, and people in those clusters would have a higher probability of becoming infected. It makes a strong case for being a hermit. The Bennett family fears social shunning when their younger daughter, Lydia, scandalously elopes with Wickham; had this transpired, they could have taken comfort in the fact that they would have been far less likely to be bitten by zombies than their more socially active neighbors.

Eagle believes that the data captured by his cell-phone software application gives a much more realistic picture of the dynamics of human social networks,* thereby arming epidemiologists with "more information to make predictions about our vulnerability to the next SARS, as well as greater insight into preventing future epidemics." One can only wonder what further insights might be gleaned if Eagle outfitted zombies with cell phones.

How human beings react to these kinds of threats is another factor that can be tough to calculate. That's why other researchers are looking to online virtual worlds for modeling the spread of infectious disease, much like the collapse of a virtual bank in Second Life might shed light on economic models. Most epidemiological models use mathematical rules to approximate human behavior, but the modelers must make certain assumptions about how humans are likely to behave—

* Ironically, the same cell phones Eagle is using to track social networks could be helping to spread one of the most virulent strains of *Staphylococcus aureus*, known as MRSA—a serious issue facing hospitals today because it is highly resistant to antibiotics. Turkish researchers tested the cell phones of many doctors and nurses in hospital operating rooms and intensive care units in Turkey. Almost 95 percent of those devices showed bacterial contamination, and only 10 percent of the staff regularly cleaned their cell phones.

and those assumptions can be inaccurate. Deliberately introducing a deadly pathogen into a controlled population to study the outcome would be immoral, but what if it were possible to design a "disease" specifically for a virtual online community?

Game designers were a little ahead of the scientists on that front. Blizzard Entertainment, the makers of *World of Warcraft*—a hugely popular multiplayer game—deliberately introduced a zombie plague into the game to promote *World of Warcraft: Wrath of the Lich King*. But a far more interesting development was the virtual Corrupted Blood epidemic that broke out in 2005. Blizzard added a new dungeon called Zul'Gurub, controlled by an "end boss" named Hakkar. Only highly advanced players could find Zul'Gurub, where the objective was to kill the end boss. Among the creature's weapons was a spell called Corrupted Blood, which inflicted damage on infected players at regular, repeating intervals, slowly draining away their vitality until their avatars "died." Killing Hakkar was the only cure.

The spell was designed to infect only nearby players, and to remain confined to the Zul'Gurub game space. But things went horribly wrong. Thanks to a glitch in the programming, the animal companions of players' avatars—technically "nonplayable"—became infected, and even though they showed no symptoms, they spread the disease to the lower levels of the game. While advanced players could survive the infection, the Corrupted Blood plague would kill a lower ranking player very quickly. Widespread panic ensued wherever the plague struck, with game spaces becoming littered with virtual corpses. At least three servers were affected, and Blizzard had to reboot the entire game to fix the glitch.

A Rutgers University scientist named Nina Fefferman heard about the Corrupted Blood incident and became fascinated by the in-game parallels to real-world epidemics. Human behavior is not necessarily rational, or courageous, and this became obvious in *World of Warcraft*. True, some players tried to help with "healing spells," but other players panicked and fled to other game spaces, carrying the disease with them. A few malicious players deliberately spread the disease—behavior that has also been documented in real-world outbreaks—and one hardy soul decided his role was to stand in the town square and narrate the carnage, a self-appointed Doomsday Prophet. There were even thrill-seekers who ignored the warnings and ventured to infected areas out of curiosity, thereby becoming infected as well—similar, says Fefferman, to journalists who travel to war zones and deliberately put themselves in harm's way to get a story. She went on to coauthor a paper with Eric Lofgren for *Lancet Infectious Diseases* on the implications of the Corrupted Blood incident for refining epidemiological models.

Fefferman's work has its naysayers, who argue that the virtual death of an avatar is not equivalent, in terms of risk, to physical death in the real world. Fefferman counters that players become quite invested in their characters and feel genuine emotional distress when those avatars are injured or killed. "The players seemed to really feel they were at risk and took the threat of infection seriously," she told *BBC News*.

Blizzard, in turn, maintains that *World of Warcraft* is just a game and was never intended to mirror reality. But the parallels to real-world outbreaks are striking. An epidemiological model based on the Corrupted Blood outbreak would draw on hard data showing how players actually responded to the threat—not on abstract mathematical assumptions. And why not look to video games for insights into the spread of diseases? We'll need all the help mathematics can give to ward off the coming zombie apocalypse. Just ask the Bennett sisters.

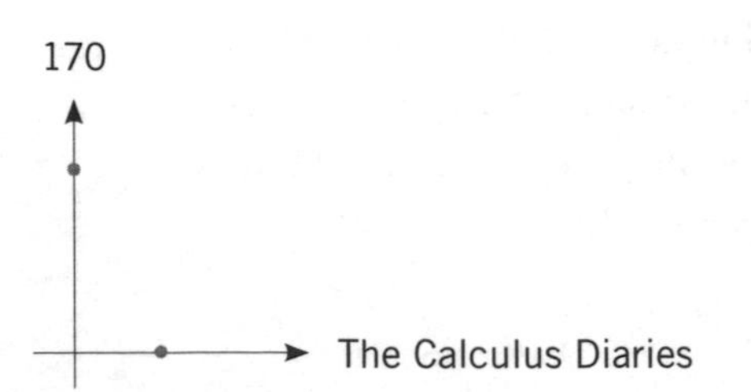

7

Body Heat

> Exercise ferments the humors, casts them into their proper channels, throws off redundancies, and helps nature in those secret distributions, without which the body cannot subsist in its vigor, nor the soul act with cheerfulness.
>
> —JOSEPH ADDISON,
> *Spectator*, July 12, 1711

Comedian Margaret Cho once riffed on the concept of "Stairmaster time"—namely, the fact that time passes much more slowly when one is on the Stairmaster, mindlessly climbing stairs to nowhere. Any gym member can relate: A mere fifteen minutes can feel like an hour if one lacks sufficient distraction. I am just beginning to break a sweat on an elliptical machine under the watchful eye of Adam Boesel, personal trainer and owner of the Green Microgym in Portland, Oregon, but we are not watching the usual graphic display showing pace, calories burned, or distance traveled. Instead, I am laboring to keep a small 60-watt light bulb alight with my exertions, mounted on the front of the machine, with just a single digital readout tracking my output in wattage.

Open since 2008, the Green Microgym is located in the Alberta Arts district of Portland, just fifteen minutes from the city airport. It's a very crunchy-granola type of place, where "sustainable" is practically a way of life. The folksy main street boasts quirky little shops, art galleries, and eateries—all owned and operated by local residents. There's nary a Starbucks in sight; instead, the local bohemians retreat to a funky little café called the Fuel Stop for their caffeine fix after yoga class, where all the drinks, salads, sandwiches, and sweets are made from scratch, and nobody minds if you stay for a few hours to take advantage of the free wireless. The hot brunch spot is the Tin Shack—a small building with aluminum siding and adjacent courtyard for outdoor dining. At 10:30 A.M. on an overcast Saturday, there is already a line of hungry locals snaking around the corner.

Boesel's Green Microgym fits perfectly in this neighborhood: a modest, two-story boxlike structure with bright red exterior and no-frills interior. There are the usual ellipticals, stationary bikes, treadmills, and free weights, but look closer and you'll notice a twist on those fitness staples. Boesel has retrofitted much of his exercise equipment so that gym members can generate a small amount of usable energy during their workouts.

He is not the first to ponder the potential of human exertion for generating energy. Inmates in nineteenth-century New York prisons were forced to walk on treadmills as punishment, and that energy was used to grind grain for the inmates' daily bread. Today, a handful of fitness centers around the world are seeking to exploit the same concept—on a purely voluntary basis. California Fitness in Hong Kong has cardio machines to produce energy for the gym's lighting, while the Netherlands boasts the Sustainable Dance Club in Rotterdam. The dance floor is made up of small modules that move in response to the people dancing, and this movement is converted into electricity that lights up the floor. A Boston gym has a special stationary bike with a laptop built into the handlebars. The laptop has no battery; it is powered entirely by the person pedaling, so someone can get in a decent workout and still surf the Web or answer a few e-mails. It is a multitasker's delight.

Back in 1990, before being energy conscious was cool, actor and environmentalist Ed Begley Jr. connected a bicycle to a 24-volt battery to generate the energy needed for small kitchen appliances. He has been known to make toast this way or to run a coffeemaker. However, Boesel and others think there might be the potential for a commercial market as well. Several companies have cropped up in recent years specializing in retrofitted exercise equipment, such as ReRev.com in St. Petersburg, Florida; Henry Works in El Paso, Texas; and entrepreneur Jim Whelan's Green Revolution. There is an entire academic research program devoted to human-powered energy at the Delft University of Technology in the Netherlands.

It's an ingenious idea. We spend hours each week running, cycling, or climbing in place, like hamsters on one of those little wheels, with no other goal than to burn off last night's indulgence in a hot fudge sundae—all in pursuit of the slim, athletic figure so prized by modern society. (Admit it: For most people, the health benefits of doing so are largely secondary.) So why not try to harness some of that energy otherwise going to waste and turn gym rats into energy generators?* The human body is essentially a machine—specifically, a heat engine. Boesel's Green Microgym is based on solid thermodynamic principles, and this makes it an ideal learning environment for exploring the calculus of energy as it relates to diet, exercise, and the economic feasibility of harvesting energy from exercise machines.

BLOWING OFF STEAM

Every March in Los Angeles' funky Echo Park neighborhood, the Los Angeles Wheelmen—a local bicycling club—gather at the foot of Fargo Street for their annual Fargo Street Hill Climb. Members compete to see who can make it up the road's steep grade between Allesandro and Alvarado Streets the most times in a single day. It's a daunting challenge: That one-block stretch of Fargo Street boasts a vertigo-inducing 32 percent grade, tying with nearby Baxter Street for the second-steepest grade in the city. (Eldred Street in the Highland Park neighborhood takes top honors with a 33 percent grade.) The current record, set in 2008, is 101 ascents, which took the stalwart cyclist, Steve Gilmore, nine hours to complete. That was an atypical year: Even the toughest Wheelmen (and -women) usually manage only between twelve and thirty climbs.

* The Swedish town of Halmstad has taken the concept of human-powered energy one step further: Residents may soon draw on excess heat from the local crematorium to stay warm in winter. Director Lennart Andersson came up with the idea after learning his facility was belching too much hot smoke into the atmosphere. "We realized that instead of all that heat just going up into the air, we could make use of it somehow," he told the London *Daily Telegraph* in 2008.

The Los Angeles Wheelmen might expend a great deal of energy biking up Fargo Street and back down again, and feel as though they've definitely gotten a good workout. But from a physicist's perspective, nothing has been done. Energy is useless unless it can be harnessed to perform some task. For example, the battery packs in the Green Microgym must be connected to some kind of load before the energy they produce can be useful—say, to operate the fans or the stereo system. Energy, when harnessed, produces *work*, which has a very specific meaning in physics, namely, a force applied over a given distance ($W = fd$). How much work a moving object is capable of performing is precisely equal to its kinetic energy.

There are many different kinds of energy that can change into each other. For example, you could also get work by burning fuel. Burning coal in power plants produces electricity by converting thermal energy (heat) into mechanical energy in a turbine. Electrical energy can change into mechanical energy. A battery relies upon a series of chemical reactions to produce an electrical current; once all the chemicals have been used up and converted into energy, the battery goes dead. And an electrical generator converts mechanical energy into electrical energy, which can then be used to power most of modern technology. All these conversions are examples of turning stored potential energy into kinetic energy.

Heat is wasted energy, for the most part, and it's the reason no machine, no matter how well designed, can ever attain 100 percent efficiency. We know this because of the work of Sadi Carnot. Born in 1796, Carnot was the son of a French aristocrat named Lazare Carnot, who was one of the most powerful men in France prior to Napoléon's ignominious defeat; the family fortunes rose and fell dramatically throughout young

Sadi's life in conjunction with that of the monarchy. Named for the Persian poet Sadi of Shiraz, Carnot learned mathematics, science, language, and music under his father's strict tutelage. At sixteen, he entered the École Polytechnique, studying under the likes of Claude-Louis Navier, Siméon Denis Poisson, and André-Marie Ampère.

It was not a peaceful period in France's history. Always opposed to the monarchy, Carnot joined in the fighting when Napoléon briefly returned from exile in 1815. When Napoléon was defeated in October of that year, Carnot's father was exiled to Germany. He never returned to France. Carnot the younger, dissatisfied with the poor prospects offered by his military career, eventually joined the General Staff Corps in Paris and pursued his academic interests on the side.

In 1821, he visited his exiled father and brother in Germany. Apparently there was very little to do in exile, so the men took to debating the pros and cons of steam engines. Steam power was already used for draining mines, forging iron, grinding grain, and weaving cloth, but the French-designed engines were not as efficient as those designed by the British. (The efficiency of those early French engines was as low as 3 percent.) Convinced that England's superior technology in this area had contributed to Napoléon's downfall and the loss of his family's prestige and fortune, Carnot threw himself into developing a robust theory for steam engines.

Carnot's father died in 1823. That same year, Carnot wrote a paper attempting to find a mathematical expression for the work produced by one kilogram of steam; it was never published. In fact, the manuscript was not discovered until 1966. In 1824, he published *Reflections on the Motive Power of Fire*, which described a theoretical "heat engine" that produced the maximum amount of work for a given amount of heat energy put into the system. The so-called Carnot cycle draws energy from temperature differences between a hot reservoir and a cold reservoir (and became the basis for modern-day refrigerators).

Carnot knew from endless experimentation that in practice, his design would always lose a small amount of energy to friction, noise, and vibration, among other factors. He knew that in order to approach the maximum efficiency in a heat engine, it would be necessary to minimize the accompanying heat losses that occurred from the conduction of heat between bodies of different temperatures. He also knew no real-world engine could achieve perfect efficiency. These considerations brought him tantalizingly close to discovering the second law of thermodynamics.

Reflections on the Motive Power of Fire did not attract much attention when it first appeared. The principle of energy conservation was fairly new and quite controversial among scientists at the time. The work began to gain notice a few years after Carnot's untimely death from cholera at the age of thirty-six, just one among the myriad casualties of the epidemic that swept through Paris in 1832. Most of his belongings and writings were buried with him, as a precautionary measure to prevent the further spread of the disease. Carnot was twenty years ahead of his time. His work did not immediately lead to more efficient steam engines, but he did set out the physical boundaries so precisely that Rudolf Clausius and William Thomson, Lord Kelvin, would draw on his work to build the foundations of modern thermodynamics in the 1840s and 1850s.

In the latter half of the nineteenth century, a British scientist named James Prescott Joule toyed with various energy sources to see which ones were most efficient. The choice of fuel can be critical, for different fuels have different conversion rates and produce different amounts of usable energy—and once again, where there is a rate of change, we're bound to find a derivative. Joule came from a long line of brewers, so chemistry was in his blood, as was scientific experimentation. He and his brother experimented with electricity by giving each other electric shocks, as well as experimenting on the servants.

Fascinated by the emerging field of thermodynamics, Joule jerry-rigged his own equipment at home (using salvaged materials) to conduct scientific experiments—specifically to test the feasibility of replacing the brewery's steam engines with the newfangled electric motor that had just been invented by measuring their conversion rates and how much useful energy they produced. It was his very own simple optimization problem.

He found that burning a pound of coal in a steam engine produced five times as much work (then known as duty) as a

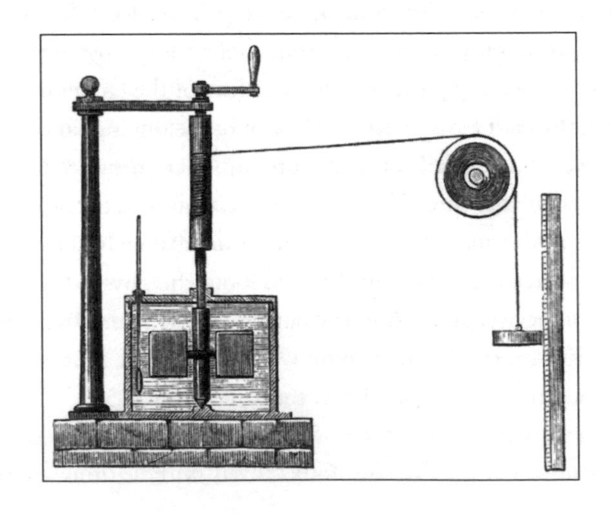

pound of zinc consumed in an early electric battery. His brewery was better off with the steam engines.

Food is another energy-dense substance, typically measured in calories. A calorie is the amount of heat energy produced when food is burned to ashes under carefully controlled laboratory conditions. It is not something that is "in" food per se. Another way to define a calorie is the amount of energy (heat) required to raise the temperature of one gram of water 1 degree Celsius (1.8 degrees Fahrenheit). The exact amount of energy required to do so is 4.18 joules. Unlike nutritionists, physicists almost never refer to energy in terms of calories. They prefer joules or watts—the derivative of joules, since watts measure the rate of energy (watts=joules per second). The "calories" in food are actually kilocalories: 1,000 calories equal 1 kilocalorie. So if I run four miles, I might burn 400 food calories (kilocalories); it sounds much more impressive when transposed into 400,000 regular calories. And that Power Bar I consume post-workout contains 270 food calories, or 27,000 regular calories—over one million joules, the unit of energy named after James Joule.

Let's see how this all applies to our intrepid Wheelman, Steve Gilmore. His body is burning food (and stored fat, assuming he has any left) for energy. But all that energy is not being harnessed for any useful purpose, other than keeping him slim and incredibly fit. He also loses a fair amount of energy as heat, radiating away a good 100 watts (8.5 million joules) every day when he's *not* riding up and down Fargo Street, and far more than that on a strenuous bike ride. His system is increasing in entropy as he is burning fuel (his breakfast). Entropy is a way to determine how much energy can be harnessed to produce useful "work." The lower the entropy, the more orderly things are, and the more useful work can be harvested as entropy increases over time. (It is a hard-and-fast law of thermodynamics that entropy always increases in a closed system.)

How much work can we get out of the system by letting entropy increase? We can determine this by taking an integral of temperature as the entropy increases over time. Gilmore's body gains potential energy as he chugs up the hill because he increases his distance from the earth's center of gravity, but he loses potential energy as he descends, while his kinetic energy increases. (Entropy also increases as he radiates away more heat with his exertions.) The two cancel each other out, and the amount of work produced is zero—unless the Wheelmen could find some way to harness the kinetic energy. For instance, Gilmore could carry a small package to the top of the hill and then work would have been done: The package has more energy as a result of his exertions. It has gained potential energy with the increase in altitude, which can be converted into kinetic energy should Gilmore then decide to toss it down the hill.

SPIN CYCLE

In the late 1980s, Henry Works founder Mike Taggett built his first retrofitted exercise bike with a car generator. He described it as a "Gilligan's Island human-powered blender" and used it to mix margaritas without an electrical outlet. Twenty years later, the same concept underlies his new retrofitted machine that makes use of both the arms and legs to maximize calories burned and watts created. It looks like your typical stationary bike, except there is a hand crank in place of the handlebars that one can spin while pedaling. All that effort turns the eighteen-inch flywheel, which in turn is connected to a generator.

Boesel started with three of Taggett's retrofitted spin bikes for his Green Microgym, and then collaborated with Henry Works to build a system of four linked Team Dynamo stationary bikes outfitted with a small motor connected to a bank of batteries. As users pedal, the motor charges the batteries, which then power the TV and stereo system. A single exerciser might only generate 50 to 100 watts of electricity—100 watts would power a small TV—but all four bikes working together can generate about four times that much, depending on how hard each person is working.

But are Green Microgyms practical? Just how much energy do all those überfit exercisers produce? Certainly Boesel's gym members are expending a great deal of effort during their workouts, but when it comes to harnessing that effort for a practical purpose, we must contend with the grim realities of thermodynamics. Boesel found that, as energy passes through the battery, some gets lost in the conversion process (entropy). So in practice, the battery pack option was less efficient in generating useful energy than a machine retrofitted with a grid-tie inverter, which sends the generated energy directly back into the power grid.

Taggett's company obliged with its FireWheel InterGrid (FIG) system. Boesel plugs the machine into a wall socket as if it were a common household appliance. The inverter is a clever device commonly used in conjunction with solar panels, enabling those who install the panels to literally spin the meter backward and sell extra power back to their local power company. In this case, the system harvests some of the ambient heat normally emitted by exercise machines (due to friction) while in operation. The battery-based Human Dynamo system uses less than 50 percent of the current coming out of the machine, while the new updated version delivers back to the grid about 70 percent of the total watts produced.

Boesel is a personal trainer by profession, and his outlook is distinctly scientific. He admits he was overly optimistic at first about how efficient his gym would be, figuring he could go off the grid completely and generate 100 percent of his energy needs. But after doing the actual experiment for nearly a year, the data told him differently, so like any good scientist, he revised his hypothesis and is seeking to refine it further. The sticking point lies in the inevitable losses that occur whenever energy changes from one form to another. Stupid entropy ruins everything.

Take a standard rowing machine. If I row furiously for ten minutes, I would burn about 100 calories. This is sufficient to run a 100-watt bulb for one hour—at least on paper. Remember that some energy is always lost in the conversion; in the case of gym rats, we lose energy by sweating off excess body heat, not to mention the enormous amount of energy required for basic bodily functions. We breathe more heavily when we exercise, and our blood circulates at a higher rate, on top of the energy required just to keep our muscles moving. So not all of the energy we generate is converted directly into useful mechanical movement. In reality, we would be fortunate to harness 50 percent of that estimated output.

The upshot is that one person on one machine simply won't

make much of a difference. Taggett estimates that one person produces about a penny's worth of electricity in an hour. But if a gym has forty retrofitted machines, all in use during the two-hour evening peak period, those exercisers would generate approximately 25 kilowatt hours of electrical energy during those two hours—equivalent to running several households for a day. This is another optimistic assessment, assuming all the exercisers are actually exerting themselves, rather than strolling on the treadmill in designer gym togs, chatting on their cell phones and not breaking a sweat. Those people bring down the overall energy output.

To maximize his savings, Boesel has combined his retrofitted machines with other energy-saving strategies. The gym has SportsArt EcoPower treadmills that run on one third less energy than traditional motors, and when machines are not in use, Boesel switches them off. The average treadmill takes between 1,500 and 2,000 watts to operate, the equivalent of nine Lance Armstrongs chugging at full power. Boesel also added solar panels to the building's exterior, and is careful not to run the A/C continuously. He has managed to keep his electricity costs to a bare minimum—about 9 kilowatt hours per month—and believes that in time, he can break even on those costs, generating 100 percent of the gym's electricity needs. At present, he figures he saves between $75 and $150 per month in electricity costs.

While Boesel estimates he can produce 75 to 80 watts consistently during his usual hour-long cardio workout, I have significantly less mass, and therefore my output is closer to 45 to 50 watts (produced continuously during the same time period), although I am not consistent: sometimes the gauge dips into the 30-watt range when I slack off the pace a bit. Seeing just how little usable energy I produce on Boesel's retrofitted elliptical is a sobering eye-opener.

Math and calculus also play a significant role in maintaining a healthy weight. The combination of how much food we eat and how much we exercise largely determines our weight, and in times of plenty, it is all too easy to consume more food than we need. Not surprisingly, human beings throughout history have devised all manner of bizarre strategies for combating their expanding girths.

BATTLE OF THE BULGE

It might be said that William the Conqueror had an overdeveloped sense of entitlement. He was the only son of Robert, Duke of Normandy, but his parents never married. His illegitimacy didn't keep him from inheriting the duchy of Normandy when his father died in 1035, on the way back from the Crusades. But he aspired to be king of England as well, having been promised the throne by King Edward the Confessor, who had no direct heirs. On his deathbed, however, Edward had a change of heart: He named Harold, son of the Duke of Essex, as his successor. Incensed, William invaded England in September 1066 and defeated the newly crowned King Harold at the Battle of Hastings. William became king of England.

William may have conquered England, but he lost the battle of the bulge. In fact, he became so fat in the years after his victory at the Battle of Hastings that King Philip of France (no doubt disgustingly svelte) cruelly described him as "looking pregnant." William was purportedly hurt, but there was truth to the statement: By that time, he could barely stay on his trusty steed. He took to staying in his rooms, subsisting on nothing but alcohol for days at a time,* but his self-designed weight loss technique failed him in the end. When William died of abdominal injuries in 1087, after falling off his horse at the Siege of Mantes, he was so fat that he barely fit into his fancy stone sarcophagus. In fact, all the pushing and shoving to get the warrior's body—horribly bloated from the heat of the day—into the coffin caused it to burst, filling the church with the stench of decay.

William the Conqueror is in good company. Excess flab (not to mention bloating) is hardly a new problem for the human race. Some modern archaeologists believe the ancient Egyptian queen Hatshepsut was quite heavy and may have been diabetic. Baseball legend Babe Ruth was notorious for his twelve-hot-dog lunches and missed much of the 1925 baseball season with what sportswriters dubbed "the bellyache heard round the world."† And U.S. President William Howard Taft infamously gained so much weight while in office that he got stuck in the White House bathtub.

In the latter days of the Roman Empire, people attending sumptuous feasts would gorge themselves on delicacies, and then repair to the vomitorium to purge their bodies of all that

* William was way ahead of his time. Nine centuries later, Robert Cameron introduced the Drinking Man's Diet in 1964. It was actually a treatise on controlling carbohydrates but emphasized that gin and vodka are low-carb libations and should be liberally enjoyed. It gave rise to the far sillier Martinis and Whipped Cream Diet.

† According to my extremely thorough copyeditor, the legend that Ruth's collapse and surgery for an abdominal abscess was *directly* caused by gorging on hot dogs apparently was invented by W. O. McGeehan. Others equally dubiously implicated a venereal disease. In retrospect, intestinal injury from bootleg Prohibition booze seems a more likely culprit. See *Babe: The Legend Comes to Life,* by Robert W. Creamer, pp. 289ff, plus the Wikipedia discussion and other sources.

excessive indulgence—back when bulimia was cool. Bingeing and purging lost its cachet as the centuries passed, and people turned to complicated fad diets to control their girth. The English Romantic poet Lord Byron struggled mightily with his weight, despite his reputation as a ladies' man (clubfoot and all). He routinely went on extreme "slimming" regimens like vinegar diets to keep his weight under control.

Around the same time, a Presbyterian minister named Sylvester Graham—one of America's earliest vegetarians—introduced the "cracker" diet, eschewing meat, rich spices, coffee, tea, tobacco, and alcohol in favor of whole-grain breads and crackers. In the early twentieth century, a San Francisco art dealer named Horace Fletcher—"the chew-chew man" or "the great masticator"—advocated controlling food consumption by chewing one's food at least thirty-two times (once for each tooth) until it was liquid, then spitting out any nonliquid residue. He lost over fifty pounds with this method, and felt one could absorb the nutrients without consuming all the calories from food.

Fad diets inevitably spawned a plethora of bestselling diet books. As early as 1727, a man named Thomas Short published *The Causes and Effects of Corpulence*, in which he advised the obese to move to arid climates, having observed (somewhat unscientifically) that heavier people tended to live near swamps. In 1864, a portly English casket maker named William Banting published his *Letter on Corpulence*, detailing how he lost fifty pounds by subsisting on lean meats, dry toast, fruit, and vegetables. It sold 58,000 copies, and the practice of dieting was known as banting for decades afterward. In 1919, Dr. Lulu Hunt Peters published another bestselling diet book, *Diet and*

Health, which introduced mass audiences to the concept of counting calories to control weight. The book sold more than 2 million copies, advocating a strict 1,200-calorie regimen.

Do you think the Aktins and South Beach diets were innovative? Think again. Back in the 1920s, William H. Hay, for example, believed proteins, starches, and fruits should be eaten separately to avoid "acidosis," claiming it "drained vitality and led to fat." He also recommended a daily enema to "flush out the poisons"—an approach that can still be seen today in the practice of colonics. Vilhjalmur Stefansson's *The Fat of the Land* praised the traditional Inuit diet of caribou, raw fish, and whale blubber, with almost no fruit, vegetables, or carbohydrates. In *Look Younger, Live Longer*, Gayelord Hauser drew the admiration of Hollywood actresses Greta Garbo and Paulette Goddard with his emphasis on vitamin B–rich foods like brewer's yeast, yogurt, wheat germ, and blackstrap molasses. He was also one of the first to develop his own line of special foods and supplements in accordance with that diet plan. Then there was the "magic pairs" diet, extolling the supposedly increased fat-burning properties of certain food combinations—lamb chops and pineapple, for example.

The twentieth century also brought the advent of diet pills and all manner of strange gadgets that were claimed to help dieters melt off the poundage while still eating whatever they liked. It all started when workers at a munitions factory in World War I inexplicably lost weight, and doctors concluded that a chemical known as dinitrophenol—used in the making of dyes, pesticides, insecticides, and explosives—was responsible for raising their metabolisms, so they burned more calories. By 1935, over a hundred thousand Americans had used diet pills made with dinitrophenol. Unfortunately, the side effects were nasty: There were several cases of blindness and a handful of deaths, and dinitrophenol was taken off the market.

My personal favorite weight-loss mechanical device is the belt-driven fat massager that wrapped around one's torso and supposedly helped jiggle fat away. It was one of many Nautilus-like machines introduced beginning in 1857 by a Swedish physician named Gustav Zander. Zander Rooms were all the rage at elite spas in the second half of the nineteenth century. Today there are Vision-Dieter Glasses, designed to make food look less appealing, and Mini-Forks to encourage diners to take smaller bites, not to mention the Diet Dam—basically a muzzle to discourage you (and those around you) from eating by making you look like Hannibal Lecter. The invention of liposuction offered a shortcut to trimming unwanted stores of fat from hips, stomach, and thighs, and in the 1950s, rumors abounded that wealthy dieters were ingesting pills containing tapeworms to help them lose weight. After dropping sixty-five pounds, opera singer Maria Callas was among those rumored to have tried the tapeworm diet, perhaps because she had a known fondness for raw steak and raw liver.

The latest technology offering new hope for expanding waistlines and flabby thighs is the free-electron laser (FEL) at Thomas Jefferson National Accelerator Facility in Virginia, affectionately known as J-Lab. FELs are useful for any number of practical applications, but back in 2006, a team of J-Lab scientists demonstrated that the laser could burn away fat in the body without scorching the top layer of protective skin. This is a very exciting development, possibly leading to revolutionary new laser therapies to treat such chronic bugbears as severe acne, artery plaque, and of course, cellulite. It offers the tantalizing possibility of a whole new way to get thinner thighs in thirty days, with no need for even a lick of exercise.

The researchers tested the concept first on actual human fat (obtained from "surgically discarded normal tissue") and then on skin-and-fat tissue samples taken from a pig. Just where did they get the pig fat for the experiment? I'm so glad you asked. Ordinarily, laboratories order their supplies from specialty outlets that cater to the tightly controlled specifications of the lab in question. In this case, for some reason, the shipping company refused to transport the pig fat the J-Lab scientists had originally ordered.

Nobody wanted to cancel the experiment, so they paid a visit to a local pig farmer. They purchased a single pig, and asked the farmer not to wash it down with vinegar—the usual custom—because vinegar would react badly with the laser. The farmer shrugged, did as he was told, and the pig met its fate. The scientists picked out a few prime pieces of pig fat and gave the rest of the pig back to the farmer. Not only did the farmer get a lot of free pork that day, he no doubt still regales his friends with the tale of those crazy scientists who paid full pig price for a few pieces of lard.

That pig died so that we might one day zap away our deposits of unsightly cellulite. But before you throw caution to the winds and order a second helping of panang curry, or an extra-large blueberry muffin with that grande mochaccino, let me emphasize that the J-Lab experiment was merely proof of concept. We are nowhere near the point where we can indulge our food cravings and burn the resulting fat away whenever we like. Operating an FEL is expensive, as is the capital expenditure required to build one. Nor is scheduling beam time at the facility as easy as scheduling an appointment with your local liposuctionist. Commercial development of any new technology takes a great deal of time and money before it can be successfully brought to market.

We are *still* looking for that magic bullet for effortless weight loss. It would be wonderful to lose weight with no muss or fuss; no need to obsessively write down in a food diary the caloric content of every morsel of food that passes one's lips; no need for specially prepared meals or supplements, elaborately orchestrated food combinations, or those telltale minute surgical scars from conventional liposuction. But there is simply no substitute for the old-fashioned method of combining a sensible diet with regular exercise to burn more calories than you consume. I guess you could call this the Thermodynamics Diet, and it has a distinct advantage over competing fad diets: It has withstood the test of time.

BURN, BABY, BURN

Lulu Hunt Peters at least had a sound scientific basis for her weight-loss approach. At the time she wrote her bestselling diet book, it had been only twenty years since chemists Wilbur Atwater and Russell Chittenden came up with the notion of measuring food as units of heat that could be produced by burning it: calories. For instance, the calories contained in five pounds of spaghetti would yield enough energy to brew a pot of coffee, while those in a single slice of cherry cheesecake would operate a light bulb for an hour and a half. If one wished to drive eighty-eight miles to visit friends or family, one would need to burn the calories contained in 217 Big Macs. (Think about that when you're planning your next road trip, and take a moment to appreciate the energy efficiency of burning fossil fuels.) If someone consumes 2,000 calories a day, that will yield just enough energy to power a 100-watt bulb for twenty-two hours—assuming 100 percent efficient conversion, which simply isn't possible, as Carnot discovered back in the nineteenth century.

Our bodies evolved into incredibly efficient heat engines, optimized for survival, and we require far fewer calories to function than we realize. The standard method for determining how many calories we need to consume each day is called the Harris-Benedict equation, first developed in 1919. It relies on estimating a person's basal metabolic rate, taking into account age, gender, height, and weight, and the resulting number is then multiplied by another number designating that person's level of activity. This would range from 1.2 for those who never exercise, to 1.9 for, say, professional athletes who exercise strenuously as much as twice a day. A 120-pound woman should consume 1,300 to 1,800 calories a day, depending on age, height, and how active she is. The average 170-pound man should consume between 1,870 and 2,550 calories a day, with the same caveats.

The Harris-Benedict equation is not a perfect method, failing to account for the fact that those with excess muscle mass will burn slightly more calories than the equation suggests, while the opposite would be true for those with excess body fat. Still, the Harris-Benedict equation can be a useful tool for weight loss. All you need to do is reduce your daily caloric intake to a number below what the equation calls for—overestimating if you are muscular and underestimating if you have excess flab. Just remember that as you lose weight, you will need fewer calories to sustain your body at that lower weight (assuming all the other factors in the Harris-Benedict equation remain the same).

Even the most chronic yo-yo dieter can recite the mantra. If you don't take in sufficient calories to give your body the energy it needs, it will begin converting fat cells into fuel—and you will lose weight. The converse is also true: If you consume more calories than your body needs, it will store that excess energy as fat. Stored fat is another fuel source for the body,* just like the food you consume. There are 3,500 calories in a pound of body fat, so it is possible to reduce one's daily caloric intake by 250 calories, burn off an extra 250 calories with daily exercise, and thus lose a pound per week.

So why is obesity so prevalent in our society? There are myriad rationales being bandied about, but from a thermodynamics standpoint, it is very simple: We are heavier than people in many other societies because we routinely consume more calories than we need for our bodies to function. This is difficult for many people to accept; they will claim they really don't eat all that much and insist they must have a slow metabolism.

---

* Beverly Hills doctor Craig Alan Bittner took the "fat as fuel" concept literally. He converted discarded fat from his liposuction patients into biodiesel for his SUV. Fat fuel yields the same mileage as regular diesel, according to the National Biodiesel Board; some start-up biofuel companies mix beef tallow and pig lard with soybean oil and other vegetable sources for their biofuels. Bittner claimed his patients volunteered their discarded fat for fuel, but several former patients filed lawsuits charging that he removed too *much* fat, leaving them disfigured. (That SUV is a gas guzzler.) He also allegedly let his girlfriend perform surgeries without a medical license. Bittner left the country in 2008 for South America.

Individual metabolic rates do indeed vary widely—and the Harris-Benedict equation takes this variation into account—as do body types, and no doubt genetics plays a role as well in determining one's natural, healthy weight.

Those arguments don't change the fundamental principle: People with lower metabolic rates need fewer calories. When they consume more calories than their bodies require—even if they eat less than "naturally" slim colleagues—they gain weight. It hardly seems fair. But who said physics was fair? Frankly, in times of famine, a low metabolism confers a distinct evolutionary advantage because it can do more with a small amount of fuel. It's when food is plentiful that this superefficiency becomes a disadvantage.*

Psychologically, we easily can trick ourselves into thinking we eat less than we really do. Studies have shown that the vast majority of us routinely underestimate how many calories we consume. (It doesn't take much to hit 2,000 calories, particularly if one is partial to fast food.) Brian Wansink is a professor at Cornell University who specializes in the study of consumer behavior and nutritional science, specifically how our environment influences our eating habits. In 2007, he and his colleague, Pierre Chandon, published the results of a study in the *Journal of Consumer Research*, demonstrating that people have become so conditioned to think that the Subway franchise's food is healthier than McDonald's that they underestimate how many calories they consume in a typical meal by as much as 21

* This is also why it's a bad idea to cut calories too drastically or to exercise excessively. The body will think there is a famine and will further slow its metabolism to conserve fuel, packing on as much extra poundage as it can. In general, losing more than two pounds per week will trigger the body's super-saver mode.

percent. Famed Subway spokesman Jared may have lost a ton of weight by eating the chain's sandwiches, but he chose the healthier options. A Subway twelve-inch Italian BMT sandwich has one third *more* calories than a McDonald's Big Mac. Wansink and Chandon also found that people tended to choose high-calorie side orders with their Subway sandwiches.

For one of his earliest research studies, Wansink focused on automatic eating patterns. People would come to the lab and eat a meal while being videotaped, then answer questions about what and how much they ate. He found that people were often unaware of second or even third helpings they consumed and denied doing so—until they were shown the videotape. Other interesting findings: People will eat 16 to 23 percent more total calories if a product is stamped with a LOW-FAT label, and switching from a twelve-inch to a ten-inch dinner plate will cause people to eat 22 percent less. All this inspired Wansink to develop his own dietary secret: "The best diet is the one you don't know you're on." In other words, small changes to the home environment and unconscious patterns can lead to big changes in your waistline.

The calories we consume are only part of the equation. At the same time, we routinely *over*estimate how many calories we burn when we exercise. The caloric numbers reported by the displays on exercise equipment feed into this misconception, because they are not always accurate, partly because they are often incorrectly calibrated and partly because when it comes to human metabolism, one size does not fit all. *New York Times* reporter Gina Kolata, author of *Rethinking Thin*, reported that while a given activity might burn an average of 100 calories per hour, for example, the range for different people could be as low as 70 or as high as 130.

Bad habits can also affect the total of calories burned. Are you one of those people who hang on to the bars while on the treadmill? You burn 40 to 50 percent fewer calories for that same activity. Do you do the same exercise routine for months at a time? As your body grows accustomed to that effort, it will need fewer calories to perform that routine. And most of the calculations used to determine the number of calories burned for various activities fail to subtract the number of calories the exerciser would be using even if they were simply sitting at home reading or watching TV.

"For moderate exercise, the type most people do, subtracting the resting metabolic rate can eliminate as much as 30 percent of the calories you think you used," Kolata writes. Even those supposedly adept at math can fall victim to self-delusion in this area. Kolata tells the story of meeting a mathematician at a conference who figured he could indulge in a slice of pie because he'd just run a quarter of a mile. "At 100 calories a mile, he might have burned 25 calories. . . . A piece of pie could easily contain 400 calories."

Personally, I adhere to the Thermodynamics Diet. The primary objective is to optimize two variables, diet and exercise, to ensure either that your weight remains constant (for maintenance) or that you steadily burn more calories than you consume so as to lose weight gradually. You don't need calculus for that, just basic arithmetic. But if it really were that simple, everyone would be slim.

First, there are economic factors at play with regard to diet: The harsh reality is that healthier foods actually cost more than junk food, so not everyone can afford a quality, well-balanced diet. Besides, some people really like pizza or French fries or a hot fudge sundae for dessert and would feel seriously deprived on a diet of lean protein, organic leafy greens, and whole grains. Surely quality of life must be factored into the equation as well. How do we find a balance?

Now calculus can be of service. In this case, we wish to maximize our "tastiness": the pleasure we derive from our food intake, given a fixed number of calories we can consume per day and a fixed amount of money we can spend on groceries. To solve the conundrum, we can plot tastiness (designated by the variable y for "yummy") as a function of diet, designated by f, for all the various foods we love that, taken together, comprise our diet. Given a diet restricted to 2,000 calories a day and a food budget of $40 per day, what small changes can we make among our current food items to maximize tastiness (y) while staying within the boundaries imposed by those two constants?

For instance, we might love Snickers bars more than brown rice and carrot sticks, but if all we ate were Snickers bars, we would quickly exceed our caloric limit, and probably develop a vitamin deficiency in the bargain. Similarly, we might love the fresh organic mixed-greens salad with free-range chicken and a light vinaigrette available at our local health-food joint, but if *that* were all we ate, we would quickly exceed our food budget. So if we know what we're eating each day now, what small change can we make in our diet to optimize how much we enjoy mealtimes?

This is a job for the derivative, with a twist. It is similar to the multivariable optimization problem we employed while house hunting, except in that case we had two variables constrained by cost; here our variables are constrained by cost and total number of calories. This makes it difficult to plot on a traditional Cartesian grid; there are simply too many dimensions to easily visualize. But we can think of it in terms of vectors, or directions of motion. There are any number of ways we can change what we eat, but some changes are not allowed because they exceed the stated limits to calories or cost; in other words, that particular vector is invalid. Other changes *are* allowed because they keep those two values fixed.

Normally we would take a derivative with respect to all possible values of f, but in this case we would take the derivative only with respect to those values for f that are allowed—namely, those that can be changed without exceeding our boundary conditions of total calories and cost.

What about the integral? It plays a role in the Thermodynamics Diet too, specifically with regard to how many calories we burn. It all comes down to the burn rate. We can take an integral of our rate of burning calories with respect to time and get the total number of calories burned—the calorie meter on an exercise machine at the gym is secretly doing this calculation. But as we've seen, that burn rate is affected by numerous variables: metabolism, level of exertion, muscle mass, and so forth, all of which complicate the equation. So most machines are incorrectly calibrated. The best those machines can manage is a ballpark figure.

THIS MORTAL CURVE

Not only is it possible to use math and calculus to optimize our diet and exercise regimen and maintain a healthy weight; we can also use it to determine the probability that we will die in any given year, thanks to the work of an obscure British actuary named Benjamin Gompertz. Gompertz hailed from a family of wealthy merchants who emigrated to England from Holland. Because he was Jewish, he was denied admission to university and thus was largely self-educated. He acquired his mathematical knowledge by reading Newton's works, among others, thereby becoming proficient at calculus.

One day, when he was just eighteen, Gompertz stopped in at a secondhand bookstore, and struck up a friendship with the bookseller, John Griffiths, who was a mathematician. Initially Gompertz wished to be tutored, but Griffiths quickly realized the young man's knowledge already outstripped his own. Instead, he introduced him to the Spitalfields Mathematical Society (later to become the London Mathematical Society), of which he was then president. Gompertz joined the Society, despite the fact that the minimum age was technically twenty-one, and found himself with more than enough math tutors at his disposal, enabling him to advance rapidly in his knowledge. (The society had a rule whereby, if a member asked another for help or information, the second member was required to provide that assistance or else be fined a penny.)

He married the daughter of another wealthy Jewish family with strong ties to the stock exchange, and that connection enabled him to join the exchange himself. He eventually became an actuary and head clerk for his brother-in-law's nascent insurance company, where his mathematical skills proved very useful. Apparently he had a great capacity for "sustained complex computation" in compiling detailed actuarial tables. In particular, Gompertz found he could apply the principles of calculus to human mortality to determine the cost of life insurance. "It is possible that death may be the consequence of

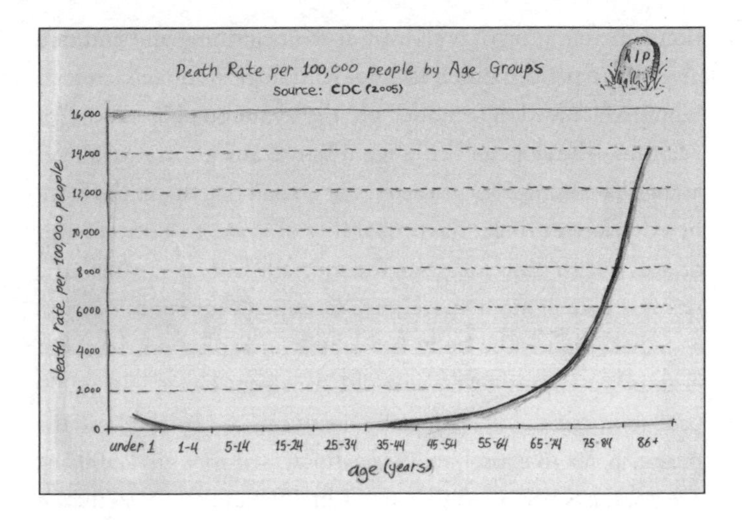

two generally co-existing causes," he wrote around 1825. "The one, chance, without previous disposition to death or deterioration; the other, a deterioration, or an increased inability to withstand destruction."

In other words, assuming one doesn't meet with a fatal accident, such as being hit by a bus, it is possible to use calculus to model the probability of the likelihood that one would die in any given year—a probability that increases with age. Gompertz tested his hypothesis by comparing the proportion of people in different age groups in four cities in England and found that mortality increases exponentially as we age. Thus was born the Gompertz law of human mortality, which holds that whatever the odds that you will die in the next year—1 in 1,000, or 1 in 10,000—those odds will be twice as large eight years from now. In other words, the probability of dying increases exponentially with time.

The Gompertz mortality curve is another sigmoid function, wherein growth is slowest at the beginning and end of a given time period, much like epidemiological models; in fact, Gompertz based his model on the demographic model of Malthus. The slope of the tangent line at any given point (age) along that curve gives us the rate of actuarial aging in the form of a derivative. To get the probability of living to a certain age, all we have to do is integrate the mortality rate over time. The result is that sigmoid curve: the "Gompertz function."

That's right: The body has a built-in expiration date. For example, a twenty-seven-year-old American has a 1 in 3,000 probability of dying during the next year, but by the time the person is 35 that probability has increased to 1 in 1,500; by age 43, it has narrowed to about 1 in 750, and so on, so that, if one reaches 100, there is only a 50 percent chance one will live to see 101. The probability that you will die during any given year doubles every eight years. It still holds true today, despite all the advances in nutrition, medicine, and quality of life, and it holds across countries, centuries, even across species, once the different rates of aging are factored into the equation. Scientists don't understand why this should be true, but it is—for the most part. There are certain age-independent factors at work as well, but in a low-mortality country, like Japan or the United States, this component is usually negligible. Gompertz himself died at the ripe old age of 86. He was working on a paper for publication in the journal of the newly formed London Mathematical Society when he suffered a paralytic seizure.

Working out and eating right to ensure better health is a noble endeavor, but sooner or later the Gompertz law of mortality kicks in. We're all going to die one day. So it is quality of life that counts, and our overall degree of happiness; being healthy increases our quality of life. Perhaps that is the real benefit of the Green Microgym concept: It might not save the planet, one workout session at a time, but it saves the gym a bit of cash and makes people feel good about their efforts, in addition to keeping them fit. "There's a certain satisfaction when you work out and feel like you've actually accomplished something, instead of just spinning your wheels," Taggett has said.

Just in case good vibes aren't enough, Boesel offers gym members special bonus points: For every hour of electricity a member produces, she or he will earn coupon points redeemable at local businesses. Most important, as I found during my own brief session on Boesel's retrofitted machine, the Green Microgym raises awareness of just how much energy we consume without thinking—and what it costs to generate that energy in the first place.

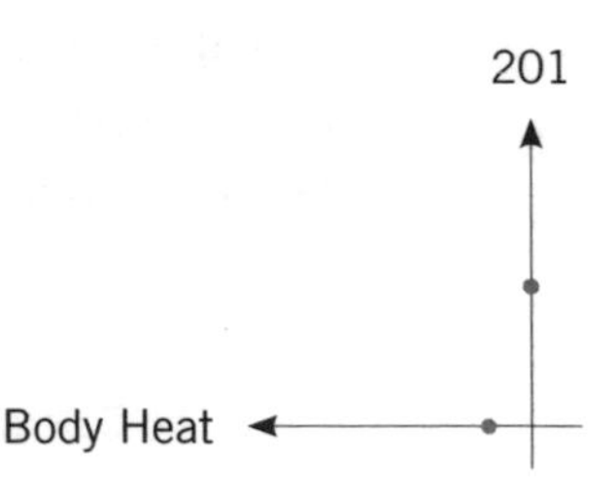

8

The Catenary Tales

As hangs the flexible chain, so but inverted
stands the rigid arch.

—ROBERT HOOKE

It is a bitterly cold February afternoon in St. Louis, the kind of day that finds most people opting to curl up by a roaring fire with a good book and a nice cup of tea. But I am only in town for a few short days while attending a conference and thus join a handful of other hardy souls visiting the famed Gateway Arch. It is a landmark structure that opened in October 1965 to commemorate Thomas Jefferson's Louisiana Purchase and now dominates the St. Louis skyline. Five of us cram into the little egg-shaped tram and ride to the peak of the arch, where we can gaze out over the frozen expanse of this Gateway to the West, named in honor of the early pioneers who migrated west through St. Louis on the first leg of the Oregon Trail.

The spectacular view is marred somewhat by our cramped conditions—the tram is a bit like a five-person ovoid coffin—and the disconcerting sensation of swaying whenever the wind picks up. The arch is designed to sway up to eighteen inches in the wind; it is closed to the public when the winds are particularly strong. For those unfamiliar with the principles of structural engineering, it may seem as if the arch were about to collapse. One woman in particular seems convinced of her imminent demise: Eyes shut tight, arms hugged tightly to her chest, she refuses her companion's entreaty to take just one look before the tram returns to the ground. A bead of sweat is visible on her upper lip as yet another gust of wind hits the arch.

The poor woman need not have worried. Appearances can be deceiving. A well-constructed arch is actually quite stable. Leonardo da Vinci once observed, "An arch consists of two weaknesses which, leaning one against the other, make a strength." The secret of how the arch stays up lies in its shape. There is a very specific geometric term for it: It's essentially an inverted model of a flexible chain or rope suspended from two points. The noninverted curving shape is known as a catenary, derived from *catena*, the Latin word for "chain." Leonardo's codependent "leaning weaknesses" describe a delicate balance of opposing forces that gives rise to a surprising degree of structural stability.

Nature has its preferred shapes. Any chain suspended between two points will come to rest in a state of pure tension, much as surface tension causes a bubble to form a sphere. In a chain, tension is the only force between consecutive links, and that force inevitably acts parallel to the chain at every point along the curve—never at right angles to it (otherwise the chain would move). Inverting the catenary into an arch reverses it into a shape of pure compression. Masonry and concrete—

standard building materials—break relatively easily under tension, but can withstand huge compressive forces. So the inverted catenary shape can be used to form structures like domes or arches that span a considerable horizontal distance.

Finnish architect Eero Saarinen didn't copy the classic inverted catenary shape perfectly when he designed this St. Louis landmark; he elongated it, thinning it out a bit toward the top to produce what one encyclopedia entry describes as "a subtle soaring effect.* But Saarinen's catenary variation is more than an aesthetic choice. He designed it in consultation with an architectural engineer named Hannskarl Bandel, who knew that the slight elongation would transfer more of the structure's weight downward, rather than outward at the base, giving the arch extra stability. He had good reason for doing so. An inverted catenary shape is extremely stable horizontally, but it is less so in the vertical direction—and the Gateway Arch rises a good 630 feet above its 630-foot base. A plaque near the site proudly declares that the Gateway Arch's distinctive shape is described by this mathematical equation: $y = 693.8597 - 68.72 \cosh(0.010033x)$.

Any engineer can tell you that math is critical to building structures with the right size, shape, and balance of forces, and that geometry plays an important role in architecture. Yet equations for geometric figures weren't even available until Fermat and Descartes devised analytic geometry in the seventeenth century, ensuring that millions of high school students would be required to take up compass and straightedge and

* Maybe not subtle enough. In 1980, a man named Kenneth Swyers took that whole "soaring effect" a bit too literally. He tried to parachute onto the arch's span and died in the attempt, garnering a posthumous Darwin Award for his effort.

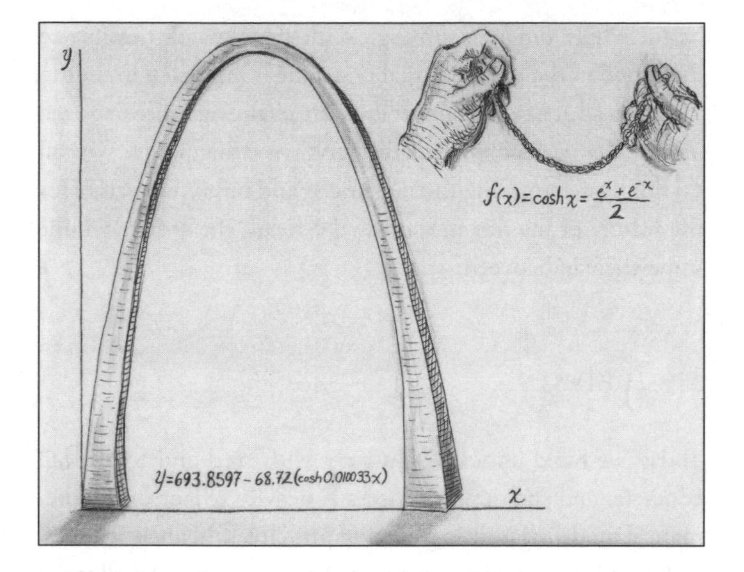

learn about interior and exterior angle sum conjectures, along with the properties of trapezoids. Today it is a relatively simple matter to calculate the volume of the famous pyramid-shaped Luxor Hotel in Las Vegas, given the precise dimensions. Wikipedia tells me that the base of the pyramid is a 556-foot square and that the structure's height is 350 feet. First we determine the area of the base by multiplying the width (556 feet) by the length (556 feet). Then we multiply that area by the height and divide that answer by 3 to get the total volume. No calculus required.

But how does one select the best possible design—one intended to optimize a particular feature, such as the optimal dimensions for a pyramid one could build given a certain amount of material, or the strongest possible shape for an arch—from a wide range of options? The tedious approach would be to painstakingly calculate each and every possible option, which would be incredibly time-consuming. With calculus, it's possible to focus not on the absolute quantities one is interested in, but to look instead at how certain features are changing relative to each other—that is, to approach the problem dynamically. We can do this by determining the maximum and minimum values for the feature of interest to narrow the focus; the answer will lie somewhere in between.

ARCH RIVALS

Today we build almost exclusively with steel and reinforced concrete, and the design process is heavily reliant on mathematical modeling and engineering principles. In ancient times, arch builders employed a method of trial and error. Small stone arches were typically built around a curved wooden form. The builder would then lay stones or bricks around that wooden form, tracing the shape with pegs and string. Legend has it that whenever an arch was constructed in ancient Rome, the architect who designed it was forced to stand underneath as the wooden supports were removed, as a means of quality control. It was a terrific motivational tool: Design it right the first time, or the arch will fall and crush you. Builders of Gothic cathedrals had to figure out how to turn stones into stable structures held together only by the forces of compression, like a stack of children's building blocks, and they did it without the benefit of analytical geometry or calculus. The oldest cathedrals have stood for a thousand years, so medieval masons clearly knew a thing or two about arch stability.

One assumes this practical knowledge was passed down through generations of builders. Yet the secret of the inverted catenary remained a mathematical mystery until the seventeenth-century English scientist Robert Hooke stumbled upon this solution to the question of an optimal shape for a stable arch. Hooke is best known for his skill at building microscopes and using them to examine the tiniest details of everyday objects, such as fleas. His exquisite drawings of what he saw through his microscopes appeared in his masterpiece, *Micrographia.* He also invented a reflecting telescope, the sextant, the wind gauge, and the wheel barometer, and he had a lifelong fascination with timepieces.

Despite these accomplishments, Hooke's stature as a scientist was largely overshadowed by that of his contemporary rival, Isaac Newton. Their professional debates over the nature of light often became intensely personal: Hooke may even have tried to block Newton's election to the Royal Society. Perhaps Hooke had cause to feel threatened: His more practical contributions to science were overlooked in favor of Newton's mathematically oriented theories. Personal vanity may also have played a role: Newton cut a distinguished, imposing figure, while Hooke was small and hunched; even his friends described him in less than flattering terms.

Hooke's pique at the lack of recognition by his peers might have been partially justified. In 2006, the long-lost handwritten minutes from meetings of the Royal Society between 1661 and 1682 were discovered wedged into a dusty nook in an old house in Hampshire, England. The manuscript laid to rest a long-standing controversy over whether Hooke or Christian Huygens had first designed a highly accurate watch with tiny spring mechanisms that eventually led to the first measurement of longitude. Hooke understood a great deal about the physics of springs, having devised the eponymous Hooke's law:

Extension is proportional to force. So when Huygens claimed to have invented a spring watch in 1675, Hooke flew into a rage, claiming someone had leaked his earlier design to the Dutch scientist. The unearthed minutes include pages from a meeting on June 23, 1670, with a description of Hooke's design for a spring watch—five years before Huygens's announcement—vindicating the homely scientist.

Interest in Hooke's contributions to science has revived in recent years, and these include a footnote in the history of calculus—specifically, the catenary and its importance to architectural arches. Thanks to his early apprenticeship to an artist, Hooke was a gifted draftsman, and his architectural bent proved useful when the Great Fire of London destroyed much of the city. It was in the process of rebuilding St. Paul's Cathedral in 1671 that Hooke "rediscovered" the secret of the catenary.

Alas, Hooke was a bit too clever for his own good. He announced his solution to the problem of the optimal shape of an arch to the Royal Society, but he never published it. Instead, four years later, he published an encrypted solution in the form of a Latin anagram in the appendix to his treatise, *Description of Helioscopes*. It did not attract much notice. Finally, in 1705, the executor of his estate published the anagram's solution: "As hangs the flexible chain, so inverted stands the rigid arch"—or, if you want to be all Latinate about it: *Ut pendet continuum flexile, sic stabit contiguum rigidum inversum.*

Had Hooke been less secretive about his discovery, he might have received credit sooner for his solution to the problem of the catenary. Instead, a German mathematician named Johann Bernoulli found the solution independently and announced it in 1691. Johann was one of eight gifted mathematicians and physicists in the legendary Bernoulli family. They were a virtual dynasty during this period. The Calvinist family originally hailed from Belgium but fled to Switzerland to escape Catholic persecution. There the family patriarch, Nicolaus, made his fortune as a spice merchant.

Nicolaus had intended that his son Johann take over the family business. Alas, Johann failed miserably as an apprentice in training and opted to study medicine at Basel University instead. In between his studies, he and his older brother, Jakob, began collaborating on the study of this shiny new mathematical tool called calculus and were among the first to apply it to various problems. Eventually, Johann switched from medicine to math, and thus began a series of nasty sibling rivalries that rippled through the Bernoulli family tree for decades.

The brothers Bernoulli were highly competitive, fought constantly—their letters to each other are filled with heated insults and strong language—and always sought to outdo each other when it came to posing mathematical challenges. (This practice of issuing challenges was all the rage back then among mathematically minded sorts.) The fact that Jakob had trained his younger brother made it difficult for him to accept Johann as an equal. Johann, in turn, hated to be outdone; he was even jealous of his own son Daniel, banning his offspring from the house when Daniel won a math contest at the University of Paris that Johann had also entered. Nor was he averse to a spot of plagiarism: He once stole one of Daniel's papers, changed the name and date, and claimed it was his own work.

That constant bickering might have been ruinous to harmonious familial relations, but it seemed to fuel the Bernoulli brothers' mathematical creativity. It was Jakob who set forth the problem of the catenary: determining the precise mathematical shape formed by a hanging chain. Nearly fifty years before, Galileo theorized that it formed a parabola, but this was disproved in 1669, leaving the matter open to debate.* Johann Bernoulli, Leibniz, and Huygens all responded with their solutions within months, beating poor Hooke to publication. In modern calculus, it is possible to find the solution of the optimal shape for a hanging chain via a minimization problem, because the goal is to minimize tension. In contrast, finding the strongest shape for an arch is a maximization problem, because we wish to find the shape with the most compression forces.

There is yet another quirky feature of the catenary: It is related both to exponential growth curves and to exponential decay curves, according to Paul Calter, a retired math professor from Vermont and author of *Squaring the Circle: Geometry in Art and Architecture*. We've already seen how both curves apply to computing compound interest, for example, or to population dynamics; the only difference is a minus sign in the relevant equation for exponential decay. Calter points out that if you fit the two curves together, you get a catenary. So in the classic catenary shape, the descending portion of the curve behaves like exponential decay, while the rising portion of the curve exhibits the characteristics of exponential growth. And this shape, when inverted, forms a stable arch.

In 1696, Johann countered with a challenge to solve a particularly knotty conundrum: the problem of the brachistochrone. The word derives from the Greek words *brachistos* ("shortest")

* In Galileo's defense, the two curves are very similar. In fact, the chains or cables on a suspension bridge initially sag in a catenary shape and then settle into a parabolic curve as additional cables are added for extra stability.

and *chronos* ("time"). Johann cheated a little, having already solved the problem himself, but the challenge was deceptively simple on the surface. Assuming two fixed points, one higher than the other, what shape would a curved path between those points have to be for a rolling ball to reach the lower point the fastest? (In fine physics tradition, this problem assumes constant gravity and ignores friction.)

You might be tempted to dredge up a bit of long-forgotten knowledge from your high school geometry class and suggest that the shortest distance between two points is a straight line—ergo, a straight line is the fastest possible path. Resist that temptation. We are dealing with a curve in this instance. Galileo proposed in 1638 that the curve would be the arc of a circle; he, too, was mistaken. If we actually performed the experiment, it would quickly become clear that the steeper the curve between the two points, the faster the ball will gain speed.

Technically, this is a minimization problem: We are at-

tempting to find the least possible amount of time it takes for the ball to descend. Yet because there is more than one quantity that is varying, the solution involves considering each and every possible path between the two points—truly a job for calculus. The solution is a cycloid, which is the curve created by a point on the rim of a wheel along a straight line.

Turn that path upside down—as with the inverted catenary curve that gives one the optimal shape for a stable arch—and you will get the path of fastest descent. You can test this result by building two tracks: one shaped like a cycloid, the other shaped like the arc of a circle, for comparison. Now roll two balls down each track simultaneously. The one on the cycloid path will reach the bottom first. Nor does it matter where one starts the ball along this curved path; it will still arrive at the bottom in precisely the same amount of time.

Five individuals solved the brachistochrone problem posed by Johann Bernoulli correctly: Johann himself; his brother Jakob; Guillaume François Antoine, Marquis de l'Hôpital; and the two founders of calculus, Leibniz and Newton. Newton was working as Master of the Mint at the time and received the challenge after a long day's work. In general, Newton was loath "to be dunned [pestered] and teased by foreigners about mathematical things." But the story goes that he was sufficiently intrigued by the problem that he stayed up until four A.M. until he solved it. All in all, it took him twelve hours. He submitted his solution anonymously to the Royal Society, but Johann was not fooled, claiming, "I recognize the lion by his print."

In the process of uncovering the solution to this puzzle, the *calculus of variations* was born; Johann's student, Leonhard Euler, refined his mentor's techniques in 1766 and coined the term. This is calculus with an infinite number of variables. One would normally try to find the optimum value for a single variable (x), but in the case of the brachistochrone problem, it is necessary to integrate over all possible curves to find the optimal solution—selecting one curve from an infinite number of possibilities.

BARCELONA'S BIZARCHITECT

Visitors to Barcelona invariably become enchanted with the city's unique architecture. In particular, one can see myriad catenary shapes in buildings designed by the great Catalan architect Antoni Gaudi y Cornet. There has never been an architect quite like Gaudi, who relied less on traditional geometric shapes and more on complicated hyperboloids and paraboloids—and of course, on the catenary. His designs also incorporate brightly colored mosaic tiles and whimsical ornamental touches like the multicolored mosaic dragon fountain at the main entrance of Parque Güell.* At least one writer has described the flamboyant Gaudi style as Gothic Psychedelia, or bizarchitecture.

Descended from a family of coppersmiths, Gaudi enrolled at the Escola Tecnica Superior d'Arquitectura in Barcelona after a two-year stint in the military. His father sold the family property to pay for his son's education, and Gaudi further earned

* A highlight of the park's main terrace is a long bench shaped like a sea serpent. Legend has it that Gaudi used the shape of buttocks left by a naked workman sitting in wet clay to design the unusual curvature of the bench's surface. Why the workman was sitting naked in wet clay to begin with remains shrouded in mystery.

his keep by working for Barcelona builders. Gaudi was not the most stellar student; he was too quirkily eccentric for that. One project involved the design of an entry gate to a cemetery. Gaudi embellished the basic blueprint by drawing a hearse and a smattering of mourners to set the mood, but forgot to draw the actual gate he'd been assigned to design. He received a failing grade. But two of his subsequent drawings received the highest marks, and eventually he earned the official title of architect. "Who knows if we have given this diploma to a nut or a genius? Time will tell," sighed Elies Rogent when he signed Gaudi's diploma in 1878.

Even today, Gaudi's work is not universally admired, and in his early career, his designs were so bizarrely original that more often than not, he was ridiculed rather than praised. (George Orwell purportedly loathed Gaudi's style when he lived in Barcelona during the Spanish Civil War.) A select few recognized the signs of genius. He soon found a patron in wealthy industrialist Eusebi Güell and began building his reputation as a rising young architect. The Gaudi of this period cut a striking figure, with his blond hair, blue eyes, and ruddy complexion—unusual for someone of Mediterranean descent. He augmented this with the most fashionable of clothes and a carefully groomed beard. In short, he was a bit of a dandy, although later in life he renounced such frivolities.

Gaudi also had a nasty temper and could be incredibly stubborn when it came to his craft. Take his design for Casa Batlló, which included every last detail, right down to the furniture. It was a truly innovative renovation, showcasing the architect's signature style, with balconies that appear to move and a large cross crowning an "undulating roof." Unfortunately, the owner of the house (Josep Batlló) had this silly notion that *his* furniture and aesthetic tastes should be taken into consideration, as he would be the actual occupant.

Their epic battle inspired a local poet and wag named Josep Carner to compose a rhyme describing a fictional "Mrs. Comes," who has been given a grand piano for her newly decorated home. Not only is its sheer size problematic, it has "no style" and disrupts the harmony of the space. Mrs. Comes asks the great Gaudi for a solution. He advises her to play the violin. Satire it may be, but the tall tale captures the essence of Gaudi: He expected others to adapt to his artistic vision, not the other way around.

It was while designing the Church of Colonia Güell on the outskirts of Barcelona that Gaudi developed his unique method for determining the best curvature for his many arches and ribs in the church's crypt, taking his inspiration from gravity. He devised a "hanging model" approach to calculate the loads on the arches. It was an elaborate system of interconnected threads, representing the columns, arches, walls, and vaults of any given design, from which he suspended sachets filled with lead shot to mimic the weight of various building components. Not surprisingly, the end result was often a catenary. Catenary shells are still used in structural engineering today.

Gaudi's method wasn't sufficient to help him design the more complex, double-curvature vaults used in the nave of his unfinished masterpiece, La Sagrada Familia, a massive cathedral still under construction 125 years later. (In fairness, 180 years passed before the famed Notre Dame cathedral in Paris was completed.) It took Gaudi 10 years to complete his design, reworking his blueprints over and over until he was satisfied with the result. His plans called for eighteen towers (twelve for the apostles, four for the evangelists, one for Mary, and one for

Jesus.) Each tower features intricate geometric designs and small ornamental sculptures, and the buttresses inside the nave look for all the world like tree trunks.

All told, Gaudi worked on the cathedral for forty-three years, twelve of them devoted exclusively to the project. For the last years of his life, he actually lived in the structure's crypt. Poor Gaudi met with an ignominious end: He was run over by a tram on June 7, 1926, while walking to the construction site, and he was so raggedly dressed that nobody recognized the famous architect. (Several taxi drivers refused to drive such a vagabond to the hospital, and were later fined by municipal police for their refusal to assist the injured man.) He wound up at a pauper's hospital, and although his friends tracked him down a day later and tried to move him to a better facility, Gaudi refused, declaring, "I belong here among the poor." He died three days later and was buried within the Sagrada Familia.

No doubt Gaudi would be gratified to learn that his masterpiece is nearing completion. Jordi Bonet, director of La Sagrada Familia since 1985, has said the interior will be completed in 2010, with plans to mark the occasion with the celebration of Mass in the main nave. After that, only one last tower must be built: the 550-foot-tall Tower of Jesus, slated for completion in 2026.* At least part of the delay was due to the fact that contractors initially couldn't figure out how to physically build

---

* Completion may be threatened by plans to build a tunnel for a high-speed rail under that portion of the city, which could cause structural damage to the cathedral, as well as two other Gaudi landmarks, Casa Batlló and Casa Milà. Such fears are not unwarranted: In 2005, a metro tunnel collapsed and wiped out an entire city block in Barcelona.

some of the bizarre structures Gaudi designed on paper. And not only did Gaudi invent his own system for calculating his catenary shapes, he did those calculations without the benefit of modern computers.

WALKING ON EGGSHELLS

John Ochsendorf remembers the first time he stood on top of the domed vault of the chapel at King's College, Cambridge. "You're standing eighty feet off the ground on a thin piece of stone," he recalls. "You can even feel small vibrations. And you can't help thinking, 'The nerve of these people!'"

Ochsendorf is a structural engineer at the Massachusetts Institute of Technology and a historian of architecture and construction. "These people" are the long-dead members of England's masonry guild who built the chapel roof around 1510. It's not difficult to see why he is so impressed with their engineering skills: The chapel's roof spans nearly 15 meters, yet it is only 10 centimeters thick—similar to an eggshell in terms of its radius to thickness ratio. "These [early arch builders] developed a very real science of construction to attain a high degree of stability," says Ochsendorf. "I'm simply in awe of the fact that we haven't surpassed it yet."

Modern architects have devised their own tricks of the trade. Like Gaudi before him, contemporary Swiss architect Heinz Isler creates what he calls reversed "hanging membranes" to design the delicate, thin-shelled dome structures for which he is justly famous. After pouring liquid plastic onto a cloth resting on a flat, solid surface, he lowers the surface, leaving the plastic-covered cloth to hang in pure tension, suspended from its corners. The plastic hardens, freezing that position. Once it has dried, Isler turns the solid shell model upside down, and that form becomes the basis for his design—a form of experimental calculus.

Ochsendorf's work is aimed at adapting a popular computer graphics tool to help unlock the elusive secrets behind the arches and domes of Gothic cathedrals. Along with his then-graduate student, Axel Kilian, Ochsendorf adapted a technique called particle-spring modeling, in which virtual "masses" at the various "nodes" of a design are connected by virtual "springs." These bounce around until they find equilibrium and are able to support the requisite loads, just like Gaudi's hanging chain. CGI animation already uses such particle-spring models to recreate the movement of fabrics and hair, because animators need to map out how forces flow in different directions in real-time 3D, and in an interactive format. Remember the scene in *Star Wars: Episode III—Revenge of the Sith* where Yoda fights an adversary while wrapped in a cloak? The movement of Yoda's cloak was designed using a particle-spring model.

Ochsendorf and Kilian realized there were parallels between the fabrics that CGI animators model and Isler's hanging membranes. A length of cloth is strong under tension, but if you push on it (compression), it simply crumples. What Ochsendorf needed was something with precisely opposite properties, so he worked out a way to turn the fabric model around. This allowed him to model architectural structures, specifically Gothic cathedrals.

He's already scored some successes with his prototype program. Ochsendorf used it to demonstrate that the domes of the

Pines Calyx conference center near Dover, England, would stay in compression under all possible loadings, thereby satisfying stringent safety regulations. Open for business since 2008, the center is topped by domes made from clay tiles glued together edge to edge. Those domes span 15 meters, yet the tiles are only 15 centimeters thick and required no supporting framework during construction. "Without Ochsendorf's program these remarkable thin-shelled shallow domes would not have been allowed to be built," Alistair Gould told me. Gould is a member of Helionix Designs, a firm based in Kent that designed the building.

Eventually Ochsendorf hopes to provide designers with a technique that could lead to revolutionary architectural designs and more environmentally friendly buildings. Many modern buildings have a severe impact on the environment. Steel corrodes with time, and the manufacture of concrete produces quantities of greenhouse gases. Ochsendorf's software program has already demonstrated that certain buildings could have been built with much less material. In essence, the program finds the solution to an optimization problem for the materials.

Take MIT's Kresge Auditorium, designed by Saarinen in 1955. It has a domed roof made of concrete 15 centimeters thick. After analyzing the geometry of the dome and feeding the measurements into his hanging-chain model, Ochsendorf reckoned that it could have been built with half the thickness of concrete, resulting in significant savings in building costs and reduced environmental impact—without sacrificing the artistry. He made similar findings about MIT's new computer science building, designed by Frank Gehry. The building features columns leaning in every direction, and the structure used roughly 30 percent more material than would have been needed if his program had been used to find where the lines of force naturally fall, Ochsendorf insists.

THE QUEEN'S GAMBIT

One of the most famous optimization problems can be found in Virgil's *Aeneid.* A Phoenician woman named Dido was forced to flee her homeland after a tyrannical brother murdered her rich husband and tried to seize her wealth. Dido didn't exactly travel light: She "fled" via several boats filled with her belongings (including her late husband's stash of gold) and numerous attendants, eventually landing on the coast of Africa, where she hoped to start a new life. Her reception by the natives was frosty at first—perhaps they'd encountered would-be colonists before—but she struck a shrewd bargain with their king, offering a substantial sum of money in return for as much land as she could mark out with the hide of an ox upon which to build her own city.

It may be that the king bought into the stereotype that women aren't inherently good at math; figuring he was getting the best of the deal, he agreed. Dido promptly took her oxhide and cut it into thin strips, which she joined together into one long strip. Using the seashore as one edge for her promised tract of land, she then laid the skin into a semicircle, thereby ensuring that said tract was significantly larger than the African king had thought possible. And on that site she founded the great city of Carthage (near modern-day Tunis), reigning as its queen. In mathematics, this is known as the isoperimetric problem: How does one enclose the maximum area within a fixed boundary?

Ah, but just how do we know that Dido's semicircle did indeed enclose the largest area given the length of that long thin strip? Calculus, of course—specifically, we must solve a maximization problem using the calculus of variations. Let's start with a simpler idealization to demonstrate the basic method. We'll assume that Dido's strip of oxhide is 600 feet long, and she wants to enclose the largest possible *rectangular* area in which the seashore provides a boundary along one side. Even in this simplified version, there are many different possible permutations she could make with that 600 feet of oxhide: long and narrow, tall and thin, and everything in between. What shape is the likely candidate for giving her the optimal square footage?

We have to start somewhere, so let's take the shape of a square as a point of reference. By definition, this means that Dido would need 200 feet of oxhide for each of the three sides, with the shoreline of the Mediterranean Sea serving as the fourth side. That gives us an area of 40,000 square feet, as area depends on length and width. However, we have no way of knowing (yet) whether this is indeed the optimal shape. So we begin varying the shape ever so slightly in different directions. For instance, if Dido arranges her oxhide to measure 201 feet down the width of two opposite sides and 198 feet across the length, she would have an area of 39,798 square feet—slightly less than a perfect square. Dido decides to test this further, and adjusts her dimensions in a different way. This time, the two opposite sides measure 199 feet in width, and the third side measures 202 feet across. The answer: 40,198 square feet. Clearly the square will not give her the most possible area.

The crucial point is that the question posed has to do with *change* in area, not simply the static values of the area—that way lies madness, for we would be randomly computing areas for different configurations in hopes of stumbling on exactly the right one. It is far more useful to consider *all* possible areas created by all possible configurations (i.e., an infinite number). We have now seen countless times how the derivative applies to any case where a change in one quantity produces a corresponding change in another quantity; the derivative measures that rate of change.

Dido can reduce the problem to a simple function: Knowing she has 600 feet of oxhide, once she chooses a width for her enclosure (w), the remaining oxhide will be evenly divided to make up the length of the plot of land. How do we translate this into an equation? We know that area equals width multiplied by length. So given the variables for width (w) and length (L) of the configuration, as well as the total amount of oxhide (600), we come up with the function $600w - 2w^2$. This is the function she would use to determine what the area would be for any given configuration she chose. Graph that function by plugging in various values for w—ranging from a width of 0, to a width of 300—and you get a pretty curve (the "face" of Dido's function). This greatly narrows the possible solutions.

Now all we need to do is find a spot along that graph where the rate of change is 0. Recall that the slope of the tangent line along a curve is equivalent to the derivative. The place where the derivative is 0 will therefore be at the very top of the curve, where the tangent line is a horizontal line and hence has no slope. And that point occurs where w = 150 feet. Ergo, Dido's

plot of land should be 150 feet wide to get the maximum possible rectangular area (45,000 square feet) on which to build the city of Carthage.

This saves Dido the trouble of having to calculate the areas of all possible rectangular shapes. She simply graphs the function and then looks to see which values give a slope (derivative) of 0. Even if there are four such spots, instead of one in this particular case, that narrows the possibilities considerably. She can certainly calculate the areas of four possible shapes to determine the best possible width for her planned city.

But remember that the optimal shape is not a rectangle, but a semicircle. To find the true optimal shape, Dido must use the calculus of variations. Just as with the brachistochrone problem, it is necessary to integrate over all possible curves—not just rectangles—to pluck the correct answer from among the infinite hordes. A semicircle with a length of 600 feet has a radius (distance from the circle's center to its arc) of 191 feet. Since we know a full circle with this radius would have an area determined by πr^2, a semicircle has an area determined by $\frac{1}{2}\pi r^2$. So a semicircle yields an area of 57,296 square feet.

Fans of Virgil know that things did not end well for Dido, queen of Carthage. She rejected the African king's offer of marriage, only to foolishly fall in love with the wily Aeneas, who wound up in Carthage with his fellow surviving Trojans after the fall of Troy. But Aeneas abandons her to fulfill his manifest destiny of founding the Roman Empire. A heartbroken Dido builds a funeral pyre, curses Aeneas, and falls on a sword given to her by her fickle lover. Aeneas and his men watch the glow of her burning pyre from their departing ship, unaware of Dido's suicide.

Later in Virgil's magnum opus, Aeneas travels to the underworld and runs into his former lover's shade, but she refuses to acknowledge him, still bitter at his abandonment. The poet T. S. Eliot once called this "the most telling snub" in Western literature. I think it shows most clearly that hell literally hath no fury like a woman scorned—particularly a formidable woman like Dido, capable of outwitting an African king with an early conceptual harbinger of calculus, centuries before Newton and Leibniz invented it.

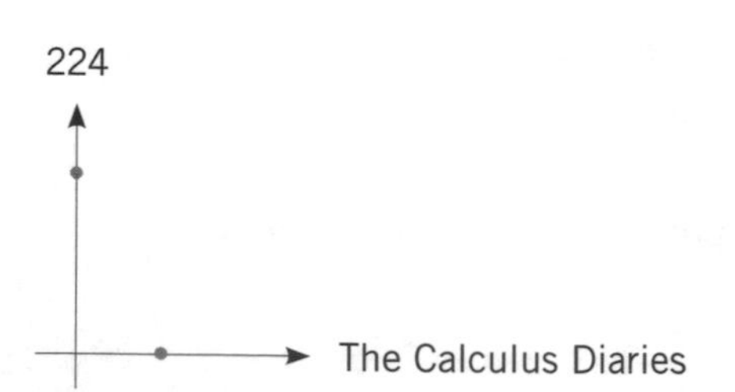

9

Surfin' Safari

> I tried surf-bathing once, but made a failure of it. I got the board placed right, and at the right moment, too; but missed the connection myself. The board struck the shore in three-quarters of a second, without any cargo, and I struck the bottom about the same time, with a couple of barrels of water in me. None but natives ever master the art of surf-bathing thoroughly.
>
> —MARK TWAIN,
> *Roughing It*

Few people are aware that novelist Samuel Clemens—better known by his nom de plume, Mark Twain—was a great admirer of the sport of surfing, or, as he called it in his travelogue *Roughing It*, surf-bathing. He even tried his own hand at surfing, with predictably dire results: He wiped out and swallowed a hefty helping of salt water for his trouble. I can empathize, as I come up sputtering from a spill for the umpteenth time during my own maiden stab at surfing in Kona, Hawaii. As I scramble back onto my beginners longboard, my self-appointed "surfing coach," Milton Garces, casually rides a swell over and calls out a snippet of helpful advice: "You might want to move back a bit on your board; you were too far forward that time to ride out the wave!"

He should know; waves of all kinds are his stock in trade, particularly sound waves and water waves. Garces is an acoustic oceanographer at the University of Hawaii–Manoa, specializing in the study of infrasound, aural signals at frequencies that lie below the range of human hearing (20 Hz to 22 kHz). Nature has an entire palette of sounds that play constantly just beyond our ken. Human hearing is rather limited in range, but sound waves exist far beyond it. We can't hear the ultrasonic shrieks of bats or the ultra-low-frequency waves of acoustic energy (infrasound) employed by elephants or tigers. Wind, water, earthquakes, avalanches, tornadoes, and hurricanes all produce infrasound, as well as audible sounds. To an acoustician, there's no such thing as perfect silence.

Most acousticians have a touch of the daredevil in them, almost by necessity: If you're trying to study the propagation of sound waves, you've got to go where the waves are happening, even if that leads you to remote Mayan ruins or the foot of an active volcano. Garces is no exception. When he's not exploding missiles at the White Sands Missile Range in New Mexico to better study the infrasonic waves that result from the explosion, he's setting up infrasound sensor arrays around volcanoes in Ecuador, or on Japan's Kyushu Island. Once he was caught napping in a Toyota Corolla in the vicinity of a volcanic eruption, resulting in some harrowing, ash-choked moments before he was able to drive to safety.

So it's not surprising that Garces is an avid surfer, along with just about everyone else in his Infrasound Laboratory (ISLA) on

Hawaii's Big Island. I have flown out to Kona to learn more about his lab, which is located right on the water's edge, the better to collect data on incoming waves. While Maui is famous for its miles of sandy white beach, Kona's shores are strewn with black lava rocks. The entire Big Island is the remnant of volcanic eruptions spanning thousands of years and is still home to active volcanoes. Locals like to place white shells against the black rocks to form pictures or spell out messages—Kona's version of graffiti.

The lab is also near at least one prime local surfing haunt; lunchtime surf outings are a common occurrence. So it seems perfectly natural when Garces insists that if I really want to understand waveforms and wave dynamics, I should experience the phenomenon firsthand by hopping on a surfboard and hitting the warm Hawaiian water. I'm a strong swimmer, and I've always wanted to try surfing, so I jump at the chance. Everyone piles into various four-wheel-drive vehicles, and we trundle our way over unpaved rocky terrain to surfing paradise.

That's how my pasty-white, city-dwelling self ends up on a borrowed surfboard in the bright sunshine, gamely paddling out to meet the incoming waves with the rest of Garces' acoustical crew, along with his wife (a scientist in her own right) and young daughter. I do not, alas, remain pasty-white. By the end of the afternoon, my entire back is bright red, even the soles of my feet. I look like a haddock that has only been seared on one side.

Sunburn aside, there is a great deal of fundamental physics involved in the sport of surfing—potential and kinetic energy, surface tension, friction, buoyancy, hydrodynamics—and in the study of waves themselves. Waves are fundamental to nearly every field of physics, from water, sound, and light, to the wave nature of elementary particles and gravitational ripples in the fabric of space-time. Not to harsh anyone's mellow or anything, but once again, wherever there is physics, there is also calculus.

BALANCING ACT

In 1778, Captain James Cook stopped off at Waimea Harbor on Kauai, en route from Tahiti to the northwest coast of North America in search of a fabled passage through that continent connecting the Pacific and Atlantic oceans. They were the first Europeans on record to visit the Polynesian chain, and their reception was warm and inviting, as they arrived smack in the middle of the season of worship for Lono, Polynesian god of peace. Islanders paddled out to where the HMS *Discovery* and *Resolution* were anchored to trade wares, so the ships could restock provisions. Cook returned after a year's fruitless searching for the Northwest Passage to restock and make repairs to his ships. But this time, he ran afoul of the natives when he stopped at the Big Island—possibly because his second landing overlapped with their season of worship for Ku, Polynesian god of war.

Historical accounts differ about the details, but it seems the conflict arose when some of the natives began pilfering items from the ships. First there was a dispute concerning a stolen pair of tongs, and then one over a stolen boat. Cook's men attempted to take a chief hostage for the return of the boat—a common leveraging practice in negotiations by British mariners—but were rebuffed. Tensions mounted, the British opened fire, and a chief named Kalimu was killed. The enraged

Hawaiians attacked in revenge, and when the British stopped firing to reload their muskets, they were driven to the water's edge at Kealakekua Bay. Cook was stabbed repeatedly with an iron dagger his crew had traded to the natives, and his body was dragged off and disemboweled, the flesh stripped from the bones. As barbaric as it sounds, it was meant as a great honor. Such were the funerary rites for the remains of a deceased high priest.

Despite the hostilities, when Lieutenant James King finally recorded the details of that ill-fated voyage in the late Captain Cook's journals, he included not just an account of the fighting, but also of the more joyous aspects of Hawaiian culture—notably surfing, "a diversion that is common upon the water, where there is a very great sea, and surf breaking on the shore. . . . They seem to feel a great pleasure in the motion which this exercise gives."

There is very little record of how surfing came to the Hawaiian islands, but by the time of Cook's visit, surfing was deeply embedded into the culture, with myths and legends about surfing heroes (and heroines) and an annual celebration called Makahiki in which surfing played a central role in honoring Lono. There were even separate reefs and beaches for royalty and commoners, a stratification that still exists in some form today: There are surf sites that cater to tourists and more-hidden local spots favored by residents.

King couldn't help admiring the skill of those eighteenth-century Hawaiians as they rode the waves, and for good reason. *La famille Garces* makes it look easy, but surfing is one of those activities that is quite straightforward in concept yet difficult to master—as Twain found out over a century ago. Following Garces' instructions, paddle a decent way out from shore, turn the board around, and wait for a promising wave. At this point, the primary physical mechanisms at work are gravity and buoyancy. (Think Archimedes and his eureka moment.) There is no acceleration and thus no net force. There is just me, on my surfboard, bobbing gently in the ocean, waiting for the perfect wave.

Whenever he spots a promising wave, Garces calls out and urges me to paddle furiously toward the shore. The trick is to accelerate to match the speed of an incoming wave just as it arrives at my position in order to "catch" it; otherwise it just shoots right past, leaving me bobbing forlornly behind on my surfboard, watching everyone else have all the fun. This happens far more often than I care to admit, due in part to my lack of upper body strength. But every now and then, I succeed and feel that telltale tug as the wave pulls me with it. At least that's what it *feels* like; from a physics standpoint, the moving wave pushes my surfboard forward, accelerating me to match its speed. At that point, I must paddle with wild abandon to ensure I end up "riding" the wave.

The first time this happens, I am so exhilarated that I throw myself off balance and promptly take a nosedive into the salty surf—a common occurrence for first-time surfers. A moving wave is literally a slippery slope, with constantly shifting forces acting on the surfboard—not just gravity and buoyancy at this point, but also hydrodynamic forces (the force exerted by a moving fluid) that push the board forward, along with a certain amount of friction or drag along the bottom of the board. You've got to keep shifting your weight back and forth to stay near the board's center of mass as you ride the wave to keep the proper balance of forces: between the downward force of gravity and the upward buoyant force. When these forces are out of balance, the board torques, or twists. If the nose is too low, you pitch forward; if you shift too far back and the nose is too high, you lose your momentum, the board stops, and you pitch into the water. In this case, the nose dipped too low, just for a second, but that was all it took: I pitched myself forward into the ocean.

It is easier to maintain that critical balance on a shorter board; the tradeoff is that it's harder to catch the initial waves. So for a beginner, like me, a longboard is best, and that is what I am using. Garces assures me the board will "catch anything" (or it would, with a better surfer wielding it); but it means it gets a bit trickier when I try to stand up once I've caught a wave. Like Twain a century before me, I wipe out on a regular basis and never quite get into a full stand; the best I can manage is a low crouch.

Hawaiian legend tells of Mamala the Surf-Rider, an Oahu chieftess who skillfully rode the biggest and roughest of waves, far from shore. I am no Mamala. Still, twice I manage to maintain my balance sufficiently to ride a baby wave all the way to the shore, with no fancy turns, but no spills, either. I'm relying on hydrodynamic forces to work their magic as water moves up the front of a wave, collides with my surfboard, and is deflected around it. If I were moving faster, there would have been a telltale spray in my wake. A good surfer—defined as "not me"—is skilled enough to keep just ahead of the break, turning up and down the face of the wave all the way into shore.

Ultimately, surfers are dancing with the waves, exploiting the same basic principles as roller coasters. They gain kinetic energy by dropping down the face of the wave and exploiting gravity, although they trade off potential energy as they lose altitude. But then they use that accumulated kinetic energy to ride back up the face of the wave to the crest, and the whole process begins all over again. Ideally, at the end of the ride, a good surfer will shift his or her weight to the back of the board, causing it to drop and the nose to rise, effectively applying "brakes." The wave rolls past, and the surfer is ready to drop back down onto the board and paddle out to catch another wave. Alternatively, you can try my cunning strategy of wiping out before I reach the shore.

That is the basic physics of surfing; where is the calculus? One simple example can be found in the knotty problem of catching that initial wave: so simple in concept, so tricky to execute. Recall that I need to reach a specific velocity—the same velocity as the traveling wave—at a specific time and place: the point at which the incoming wave reaches me bobbing in the water on my borrowed surfboard. A baseball outfielder merely has to be in the right place (position) at the right time; a surfer must match velocity as well. From a calculus standpoint, it's a matter of integrating acceleration over time in order to hit the matching velocity at precisely the moment the wave reaches me. Technically, we have to take two separate integrals—one to determine velocity by integrating over acceleration, and another to determine position by integrating over velocity—to ensure I catch that incoming wave.

"Really, it's amazing that anyone can possibly surf at all," Sean observes as he ponders the mathematical realities of the sport. And yet excellent surfers abound, every last one of them a master at making that intricate calculation within seconds, many without consciously realizing they are doing so. The human brain is capable of performing amazing feats of calculation, although this is as much a learned as an innate ability. When it comes to sports and motor skills—and calculus, for that matter—practice makes perfect.

WHAT'S YOUR SINE?

Surfing entered the international mainstream in 1959 when the film *Gidget* hit the silver screen, coining the term "the Big Kahuna" to describe the best surfer on the beach. Traditionally, a *kahuna* was a local priest or magician who would intone special chants to christen new surfboards and bring promising surf conditions. In reality, the size and shape of ocean waves depends not on mystic chants, but on three variables: wind speed, the "fetch" (the distance of open water the wind has been blowing over to form the waves), and how long the wind has been blowing over a given area. The best waves, according to experienced surfers, are those produced by intense distant storms that generate heavy winds. Those winds blow continuously for several days, creating lots of waves that slam into each other repeatedly to create a "chop." Gradually, all the little waves accumulate into a larger swell. By the time they reach the shores of Hawaii, they've become a series of powerful, large swells.

It is not a big-wave day when I have my outing—good news for me, as a beginner, because the waves are smaller and the waters less crowded with hard-core surfers. One of the keys to surfing is choosing the right incoming wave. This is not an easy call to make; ocean wave dynamics are pretty complex.

That's why surf forecasters rely on real-time meteorological data from satellites to locate the biggest waves. Avid surfers get pretty adept at eyeballing the incoming waves to identify the most promising (by size, by when they're likely to break, and so forth) and also at estimating how fast those waves will be traveling by the time they reach the surfer. But to a novice like me, they all look the same, and it's tough to predict when they'll crest and break.

Calculus comes into play when analyzing the waves themselves. All types of waves have three basic properties: wavelength, frequency, and amplitude. In the case of sound waves, the distance between compressions determines the wavelength. Objects that vibrate very quickly create short wavelengths because there is very little space between the compressions, creating a high-pitched sound. Objects that vibrate very slowly create long wavelengths because the compressions are spaced further apart. Frequency measures how many crests, or compressions, occur within one second; the measurement of this speed of vibration is called a Hertz (Hz), and 1 Hz is equivalent to 1 vibration per second. A sound wave's amplitude, or range of movement, determines the volume (loudness) of the sound.

Ocean waves have these properties, too. Generally, waves are measured according to height (measured from trough to crest), wavelength (measured from crest to crest), and period (the interval between the arrival of consecutive crests at a fixed point), corresponding in turn to amplitude, wavelength, and frequency. Mathematically, waves are described as periodic functions: They repeat over regular intervals, forming a series of crests and troughs over time. Graph a periodic function on a Cartesian grid, and you get the signature ideal sine wave, the "face" of a periodic function:

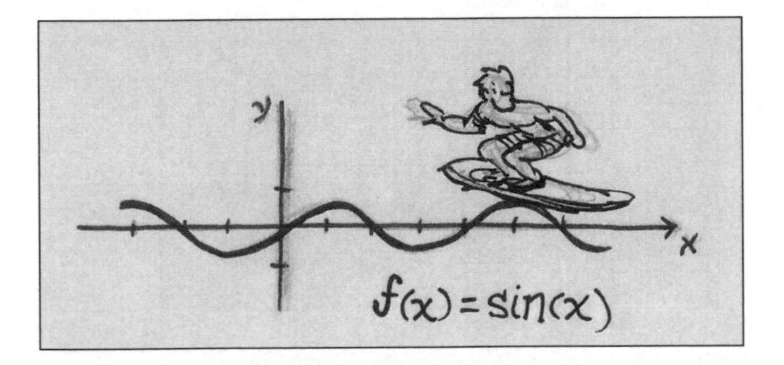

The above sine wave represents the function sin(x), the mathematical idealization of a wave. The cosine, or cos(x), is the complement of the sine. It looks very much the same, except everything is shifted slightly to the left along the x axis, such that the cosine wave appears to start at its maximum while the sine wave starts at zero on the graph.

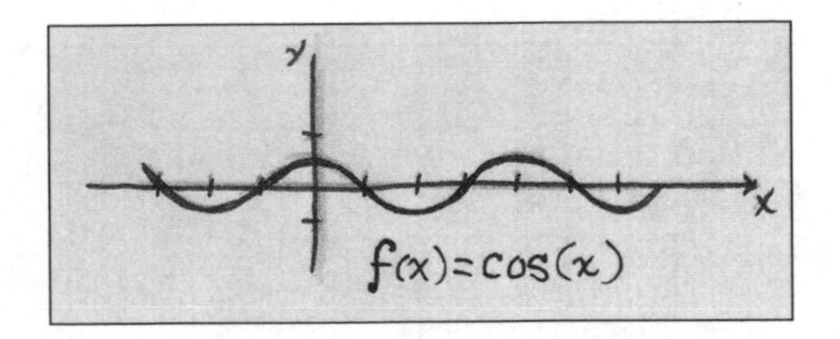

Any wave can be thought of as a sine or cosine, merely shifted by different amounts. Mathematicians simply call such waves *sinusoids*. We can glean quite a bit of information about a waveform from its "face." The number of crests and troughs we can count in a given period of time, such as one second, gives us the frequency of the sinusoid. A large number of crests and troughs means it is a high-frequency wave; a low number of

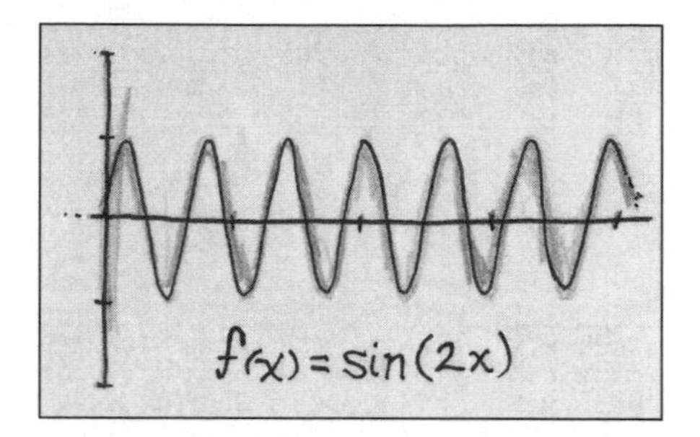

crests and troughs means it is a low-frequency wave. If we multiply x by a number, like 2, we increase the frequency of the wave (above), described by the function $\sin(2x)$. This means that the periods between crests and troughs will be shorter, giving us a wave with a higher frequency.

We can also adjust the amplitude—how strong/loud the wave is—by multiplying the function by another number, such

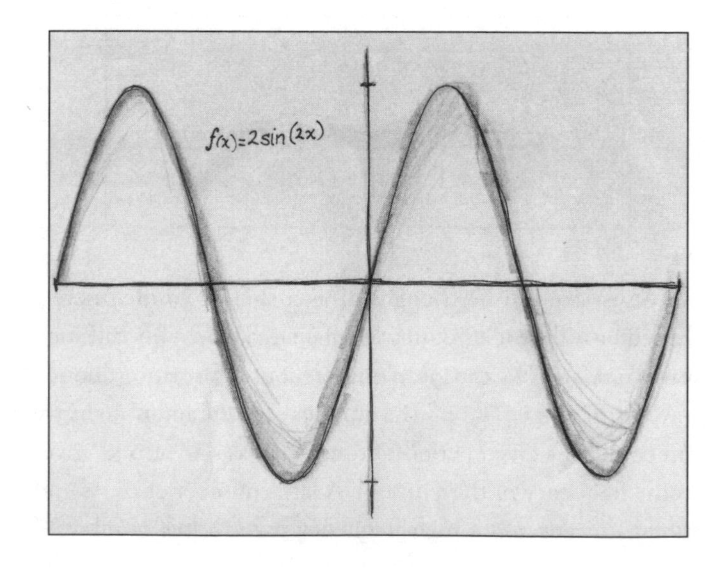

as 2, giving us the function $2\sin(2x)$. The resulting graph on page 236 shows a sine wave with higher crests.

The sine and cosine are the simplest waveforms, equivalent to pure musical notes, or a light wave of a single color. Different kinds of waves can interfere with each other, mixing together to form more complex waveforms. Sines and cosines can be treated just like any other function in calculus; only the notation is different. We can still take derivatives and integrals, and those values correspond, respectively, to the slope of the tangent line and the area under the curve. There are some interesting connections between sines and cosines that provide shortcuts when taking derivatives and integrals. For instance, note that the sine wave starts at 0 on the graph, rises, and flattens out at the peak; conversely, the slope of the tangent line of our sine wave starts at 1 and goes to 0. This is exactly how the cosine wave behaves, and we can deduce from this that the derivative of the sine is the cosine.*

Similarly, the cosine wave starts at 1 and goes to 0, while its slope starts at 0 and goes to minus 1 on the graph. So, the derivative of the cosine is minus the sine. Working our way full circle, we see that the same holds true when we're talking about finding the derivatives of minus the sine and minus the cosine: the derivative of minus the sine is minus the cosine, while the derivative of minus the cosine brings us right back to where we started: the sine.

The integral follows the same circular pattern in reverse, as it undoes the work of the derivative. The integral of the sine is minus the cosine; the integral of minus the cosine is minus

* Technically, this is true only if we are measuring all angles (x) in radians, as opposed to degrees. One radian is equal to $180/\pi$ degrees.

the sine; the integral of minus the sine is the cosine; and the integral of the cosine is the sine. The above holds true whether we are talking about sound waves, light waves, gravitational waves, or ocean waves. So we can use calculus to analyze any kind of change and motion in wave phenomena.

BREAKING THE WAVES

Most ocean waves eventually "break" as they move into shallower water, which is what happens when the wave base can no longer support its top, causing it to collapse. On this Kona beach, the waves don't break all at once, but peel to the right or left when they break. We have the pleasant spilling or rolling version of breaking waves. The plunging variety can break too suddenly, dumping surfers and pushing them to the bottom with a lot more force than one might think. There's a lot of energy in those ocean waves: Depending on the size, it can be as much as five to ten tons per square yard. Surging waves might not even break, but their powerful undertows can drag unwary swimmers and surfers into deeper, more dangerous waters.

Breaking waves produce infrasonic signals as well as audible sounds, and Garces' work exploits this feature to develop a technique he calls real-time surf infrasonic monitoring, or, as he describes it, "the deep sound of one wave plunging." Garces is specifically studying breaking waves along Oahu's North Shore, widely deemed to be a surfer's Mecca.

There are three types of wave breaks that produce infrasound: plunging breaks, cliff breaks, and reef breaks. Garces' research focuses on the latter. He is attempting to isolate the sound of a single wave in the process of breaking. Essentially, he's tracking moving wavefronts with sound sensitive pressure sensors strewn along the ocean floor, enhanced with conventional seismography. The idea is to use the collected raw data to determine wave height and other properties, for example, to better identify potential hazards to surfers. It's trickier than it seems: Such predictions currently rely on the observations of surfers themselves to determine wave heights. True, there are sensor-equipped buoys in the cove designed to collect that information, but the data are insufficient to make accurate predictions.

This might seem surprising, since a similar buoy system works quite well along the coastline of San Diego, where the Scripps Institute deploys a set of buoys and crunches the raw data using clever algorithms to separate the meaningful signals from background noise. This enables them to plot the direction, speed, and curvature of incoming waves to determine the location of the sound source and to make more accurate predictions.

So why wouldn't it work on Oahu? I asked Geoffrey Edelmann, an acoustician at the Naval Research Laboratory, who explained that it's easier to establish directionality along San Diego's far more sheltered coastline than it is in Hawaii, where wave directionality isn't clear at all—the waves are literally coming in from all directions at once. So the San Diego algorithms don't apply; scientists can't make the same set of underlying assumptions. But if Garces' hunch turns out to be right, infrasound could end up being a very useful tool for oceanographic monitoring in that region.

The raw infrasound data Garces collects requires a great deal of signal processing and analysis before real-time surf infrasonic monitoring can yield useful insights into ocean wavefronts. Because the waves that eventually hit the shores of Kona are an accumulation of many different waves of varying frequency, part of that processing involves breaking down complex waveforms into the individual component waves. This can be done thanks to a method devised by eighteenth-century French mathematician Jean-Baptiste-Joseph Fourier, a procedure known as a Fourier transform. I wasn't able to find any historical evidence that Fourier, like Twain, ever tried his hand at surfing. But I'm sure Fourier would have made an excellent surfer—at least in theory. This was a man well versed in periodic functions.

Fourier had a gift for making waves. Born in 1768, he was the son of a very fertile tailor in the village of Auxerre; Fourier had eleven siblings, as well as three half-siblings from his father's first marriage. Orphaned by age ten, the young Jean-Baptiste received an early rudimentary education at a local convent, thanks to a recommendation by the Bishop of Auxerre, and he proved such an apt pupil that he went on to study at the École Royale Militaire of Auxerre. There he fell in love with mathematics, although he initially planned to enter the priesthood. Math won out in the end; by 1790 Fourier was teaching at his alma mater in Auxerre.

Perhaps his desire to focus on mathematical research—and his inability to accomplish much of significance in his earliest years—was influenced by the tumultuous times in which he lived. Revolution was brewing in France. Fourier was sympathetic at first to the revolutionary cause, drawn by "the natural ideas of equality," and a hope "of establishing among us a free government exempt from kings and priests." He joined his local Revolutionary Committee but soon regretted it, as the ultraviolent Reign of Terror gripped France and thousands of nobles and intellectuals fell victim to the guillotine. The streets of Paris literally ran with blood.

It was frighteningly easy to run afoul of the murderous mob mentality that prevailed during the Terror; the movement soon splintered into squabbling factions, despite sharing similar goals, and rampant hysteria spread throughout France. Wise men kept their heads down and tried not to attract attention, as almost anyone could be accused of treason for the slightest perceived infraction against the new republic in that volatile environment. Fourier made the mistake of defending the stance of his own Auxerre faction before a rival sect while on a trip to Orléans. In July 1794, he was arrested and imprisoned for the views he'd expressed on that trip, and found himself facing the guillotine.

He was fortunate that his imprisonment occurred just before Maximilien Robespierre—mastermind of the Reign of Terror—ran afoul of his own revolution and lost his head to the angry mob he helped incite. With the death of Robespierre, the Revolution lost steam, and Fourier and his fellow prisoners were freed. Fourier had the good fortune to be selected for a new teacher-training school to help rebuild France, where he studied under three of the most prominent French mathematicians: Lagrange, Laplace (who wisely fled Paris during the Terror), and Gaspard Monge. By September 1795, Fourier was teaching at the prestigious École Polytechnique.

All this occurred before Fourier turned thirty. But the quiet life of contemplation still eluded him. A few years after his academic appointment, he joined Napoléon's army as a scientific advisor when Napoléon invaded Egypt. Mostly he engaged in archaeological expeditions and helped found the educational Cairo Institute, as Napoléon's military fortunes in Egypt waxed and waned. By 1801, Fourier was back in France, teaching, until Napoléon whimsically appointed him prefect in Grenoble. At long last, Fourier was in a stable, peaceful environment where he could focus on mathematics—and he promptly stirred up a mathematical controversy.

MIXING AND MATCHING

The culprit was a single equation describing how heat traveled through certain materials as a wave. Fourier concluded that every wavelike "signal," no matter how complex, could be rebuilt from scratch by adding together many different waves mixed together according to a specific "recipe." In other words, complicated periodic functions can be written as the sum of simple waves mathematically represented by sines and cosines (now known as the Fourier series). We can figure out which waves are present in a complex signal by taking an integral over all possible waves. That is the Fourier transform.

Fourier transforms are difficult for a beginning calculus student to grasp, and more complex signals require powerful computers to crunch the numbers, but the overall concept is straightforward enough. You just take apart the original signal to determine the "ingredients," and then figure out how to rebuild that signal with a mixture of the same component sinusoid waves.

It's a bit like trying to re-create at home your favorite restaurant's *spécialité de la maison*, except you have to guess at the ingredients. The more sinusoids we use, the more accurately the resulting reconstructed waveform resembles the original—much as estimating the area underneath a curve gives a more accurate result if you use more and more rectangles in the method of exhaustion. Anytime we are adding together many different smaller pieces that add up over time, we are taking an integral.

There is a neat trick to determining whether a given waveform is an ingredient in our original signal. Earlier we saw two simple sine waveforms, representing the functions sin(x) and sin($2x$). If we multiply sin(x) by itself, we get a wave that looks like this:

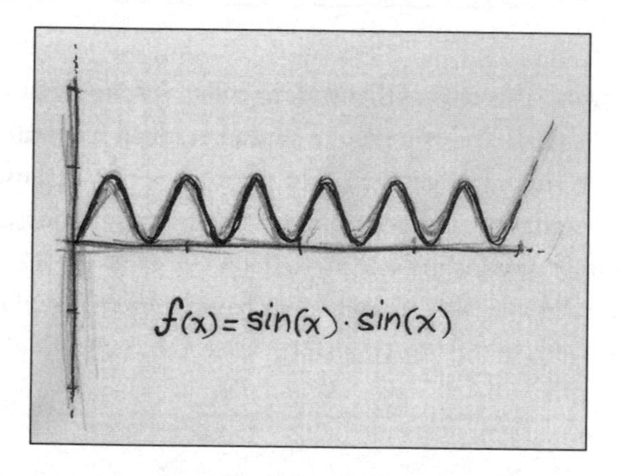

Note that it oscillates entirely above the x axis, unlike the original sine wave, which oscillated equally above and below the x axis. If we integrated it, the total area would gradually accumulate; it would just go up and up, with no subtractions. This is how we know that sin(x) is a component of our original signal—indeed, it is the only component wave of our original signal. In contrast, if we multiply sin(x) by sin($2x$), we get a resulting wave that looks like the graph at the top of page 244.

This time, it oscillates fairly equally above and below the x axis. If we integrated it, the total area would oscillate around 0, because sometimes the area adds to the total, and sometimes it subtracts. This tells us that sin(2x) is *not* a component of our original signal. We would get a similar result if we multiplied sin(x) by sin(1.1x), sin(3x), or any other wave, because our original signal was not a complex waveform, but consisted of one simple wave as the sole ingredient.

Let's see what happens when we have a signal that adds two waves together: the function sin(x) + sin(2x), which looks like this:

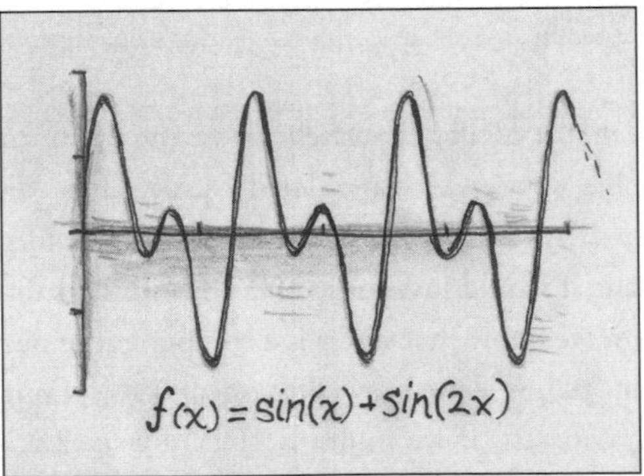

Now we perform the exact same process for each possible sine wave that could be a component. For instance, multiply the above wave by sin(x), and we get this:

$f(x) = \sin(x) + \sin(2x) \cdot \sin(x)$

Since most of the oscillation occurs above the *x* axis, we know that if we integrated it, the total area would accumulate, indicating that sin(*x*) is one of the components of our original signal. But if we multiply our original signal by sin(3*x*), we get something that looks like the illustration on the illustration below.

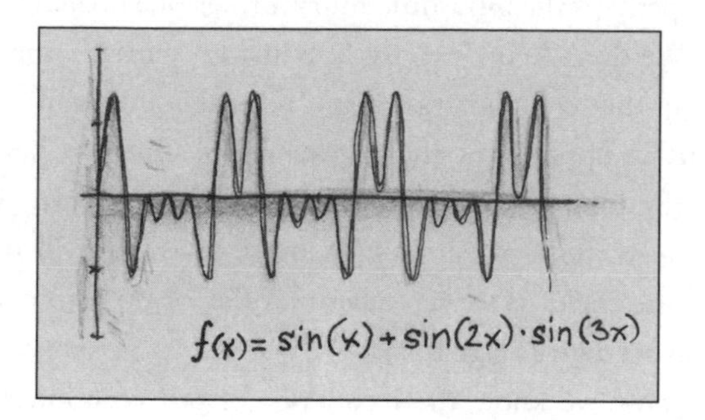

This tells us that sin(3x) is not a component of our original signal, because it oscillates equally above and below the *x* axis. We would get a similar result for every other possible wave we tried—a variable that is commonly designated in an equation by the Greek letter omega (ω)—that was not either sin(*x*) or sin(2x), because those are the only two ingredients of our original waveform. Ultimately, what the Fourier transform does is turn our original function of x into a function of ω. The integral is what makes that transformation complete.

Digital signal processing (DSP) would not be possible without Fourier analysis. DSPs are microelectronic devices that determine which sound wave is required to cancel noise. A DSP contains a resonator that vibrates in response to specific incoming frequencies. It then re-creates that sound wave—minus the frequency it is trying to cancel—and amplifies it through speakers or headphones. The end result is near silence. Most cell phones, CD players, and hearing aids now contain one or more DSP devices. The Fourier transform is a mathematical resonator, an efficient tool to filter signals.

A DSP first selects a sampling of a given signal measured at regular intervals. Let's say we have 1,000 such samples of that signal, perhaps the infrasonic murmurings of a breaking ocean wave. We don't know exactly how many "partial" sine waves make up that complex waveform, but the number of samples gives us an upper limit and a lower limit. Once we have that range of values, we can determine how many sine waves would fit between those two limits. In most cases, we will need as many sine waves as we have samples (i.e., 1,000) to perfectly reconstruct a given signal.

So now we know the frequency of our component sine waves. We also need to know their amplitude, that is, how much of each partial sine wave to mix in as we rebuild our original signal. If a given signal is heavy on the bass, there are more low-frequency sine waves mixed in than high-frequency ones, and vice versa if the signal is shrill with more treble in the mix. How can we figure that out? Why, by taking an integral, of course. We multiply that sine wave with signal samples and add the results together to get the amplitude of the partial sine wave at the frequency of interest. We do this for every possible frequency of sine wave, weed out those that we identify as ingredients, and voilà! We end up with the recipe for rebuilding that signal.

That evening finds me unwinding from the day's exertions over a refreshing tropical cocktail in the hotel's outdoor bar. The sun is setting, casting a pinkish orange glow over the water as hotel staff readies for the night's luau performance. I find myself listening to the rhythmic cycle of waves crashing on the shore and reflecting on the fact that I can hear those waves because my brain is performing Fourier transforms constantly. It's an essential part of how we hear.

The brain senses the incoming pressure wave and performs a Fourier transform on that signal to identify the frequency and amplitude of the "sound." The ear measures change in pressure as a function of time. Sometimes it is a single note; sometimes it is several notes together, as in a musical chord; in each case, the brain uses the Fourier transform to determine which components make up the total sound wave. Similarly, every time we gaze at a sunset and identify specific hues—a spectacular orange-red, or a more subtle pinkish glow—our brain has taken a Fourier transform to isolate specific frequencies of light. And when a surfer eyeballs incoming ocean waves, his or her brain is making similar calculations.

The Fourier transform has a personal significance as well. Shortly after becoming engaged, Sean and I drove from a conference in San Francisco to our new home in Los Angeles via the scenic route along the Pacific Coast Highway. At sunset, we stopped briefly to refuel just north of Malibu and found ourselves admiring the brilliant orange, red, and purple hues stretching across the darkening horizon, savoring the peaceful sound of waves lapping against the shore. It was the perfect romantic setting to cap off a long day's drive. Sean is nothing if not romantic. He is also the quintessential physicist. So he put his arms around me and whispered, "Wouldn't it be fascinating to take a Fourier transform of those waves?"

I will never listen to ocean waves or view a beautiful sunset in quite the same way again. That is perhaps the greatest gift one can gain by delving into calculus: It is a whole new way of looking at the world, accessible only through the realm of mathematics. I looked out over the ocean that evening and saw a picture-perfect ocean sunset, but there was so much more that I missed. Sean looked out onto the same scene and saw the rich complexity of nature expressed in mathematical symbols, the fundamental abstract order lying just beneath the surface.

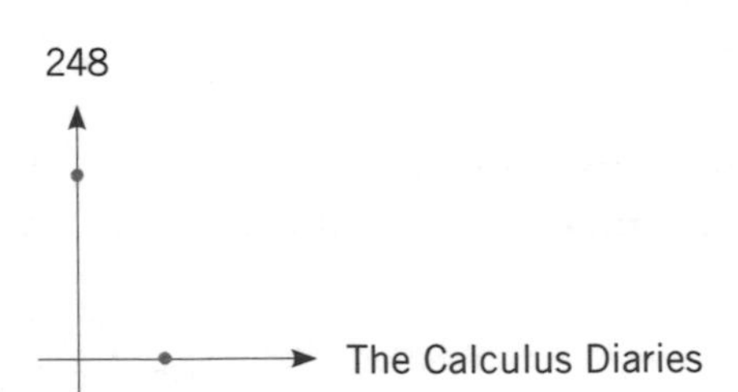

EPILOGUE

The Mimetics of Math

> I'm very good at integral and differential calculus,
> I know the scientific names of beings animalculous;
> In short, in matters vegetable, animal, and mineral,
> I am the very model of the modern Major General.
>
> —GILBERT AND SULLIVAN,
> *The Pirates of Penzance*

A pretty peach-hued building with an octagonal turret facing the Pacific Ocean is nestled on the edge of the University of California campus in Santa Barbara. This is the Kavli Institute of Theoretical Physics, where the world's best physicists gather to exchange ideas that will usher in the revolutionary breakthroughs of tomorrow. The setting is idyllic, right next to the beach, so the siren song of sun and surf inevitably vies for my attention. On this particular day, the science is winning. There is a "blackboard lunch talk" by Joe Burns, a friendly and engaging astrophysicist from Cornell University. He is among the many scientists involved with analyzing data collected by NASA's Cassini spacecraft orbiting Saturn to learn more about this distant planet—especially its mysterious rings.

It has been my custom during technical talks at KITP to focus on the concepts and let my eyes glaze over whenever an equation appears; much of the math is far too advanced for a fledgling calculus student to follow anyway. At first, Burns's talk—while less technical than some of the brain-melting lectures I've attended—looks to be no exception. Inevitably, Burns turns to the blackboard and starts scratching out equations. But this time, I recognize the notation. Burns is taking a derivative. In a flash, I realize this means he is calculating a varying rate of change: the minute changes in velocity of the millions of icy particles (ranging from the size of seashells to surfboards) that orbit Saturn and make up its rings.

That lecture was a "mimetic moment" for me—the point where the abstract symbols in my calculus books finally began to make some sense, because I could connect them with something recognizable in the real world. In ancient Greece, *mimesis* referred to the artistic representation of nature, although two philosophers differed dramatically in their interpretations of the term. In one corner, I give you Plato, of cave-allegory fame, who believed in a divine realm of Ideal Forms. All creation, including Nature, was imitation in his eyes, and artistic imitation was by definition twice-removed from the Ideal. Ergo, all art (created fictions) is inferior to the "real" world, which is in turn inferior to the realm of Ideal Forms.

In the other corner, we have Aristotle, who took some time off from speculating that we see by shooting rays of light out of our eyes that reflect off nearby objects, to write his famed treatise *Poetics*. Aristotle was more forgiving of mimetic make-believe, for he thought that human beings have an inherent need to create artistic fictions as a form of catharsis, although he valued tragedy over comedy, via a rather convoluted process of reasoning. (He was wrong about how human vision works, too.) Our modern aesthetic still owes something to Plato and Aristotle, both of whom distinguished between *diegesis,* the act of telling, such as indirect narration of action or lecturing to students about calculus, and mimesis, the act of showing a character's internal thoughts and emotions via external actions. It's a dictum of modern entertainment: Show, don't tell.

Anyone who's taken Philosophy 101 could tell you that much. But in 1946, a literary scholar named Erich Auerbach adapted the concept of mimesis in what his biography at Lerhaus.org claims is "one of the most ambitious works of literary theory ever undertaken." *Mimesis: The Representation of Reality in Western Literature* is pretty much required reading for serious students of literature; it had a profound effect on my undergraduate self, and a copy still graces my bookshelves. Auerbach analyzes literary conventions throughout the history of Western Europe and how they create "a lifelike illusion of some 'real' world outside the text." My college English professor described the mimetic moment as the point at which one makes the critical connection between one's own experiences and the artistic work and realizes, "Aha! *This* is *that*!" This kind of emotional and intellectual resonance on the part of the audience is what makes the creative arts so powerful.

I've spoken to many a scientist who was inclined to agree with Plato in devaluing fiction, which is a shame, because I would argue that created fictions present a uniquely effective teaching tool, a way to supplement rather dry college lectures (diegesis) with a dose of creativity (mimesis) to spark students' excitement and interest.* Show; don't just tell. The mimetic moment is a critical component of acquiring true knowledge—actual learning, as opposed to memorizing facts by rote. Learning science, math, or any other subject is all about making that critical connection.

While I was at KITP, I got to know mathematician Bisi Agboola, who teaches at UCSB. Bisi was educated in the United Kingdom and failed most of his math classes through their equivalent of high school: "I found it dull, confusing, and difficult." As a child, he was determined to find a career in which he wouldn't need any math, finally announcing to his skeptical parents that he would be a woodcutter. He was crushed when they pointed out that he would need to measure the wood.

But one summer he encountered a Time-Life book—simply titled *Mathematics*, by David Bergamini—on the history of mathematics, from the Babylonians up until the 1960s. "It captured my imagination and made the subject come alive to me for the very first time," he said, and it changed his mind about this seemingly dry subject. He realized there was beauty in it, and he wound up teaching himself calculus. Today he is a mathematician specializing in number theory and exotic multidimensional topologies. But he still doesn't much like basic arithmetic: "I find it boring."

Different people learn in different ways. Some students respond well to how calculus is traditionally taught, while others, like Bisi (and me), don't; but that doesn't mean we lack the aptitude to learn. That was the viewpoint of an eighteenth-

* It's been said by more than one educator that physicists use fictions all the time in the classroom, since standard introductory textbooks ignore complicating factors such as friction.

century educational pioneer named Johann Pestalozzi, whose ideas laid the groundwork for modern elementary education. Born in Zurich, Switzerland, Pestalozzi was the son of a physician who died when Johann was quite young. He was raised by his mother and grandfather in a rural village, and that experience gave him a lifelong empathy for the plight of the Swiss peasantry. While at university, he embraced the "natural" philosophy of Jean-Jacques Rousseau, even naming his son in the great thinker's honor, and went on to become a schoolteacher.

Many of his ideas were quite radical. Pestalozzi rejected the "tyranny of method and correctness" that pervaded Swiss schools of that era, declaring that he wished "to wrest education from the outworn order of doddering old teaching hacks as well as from the new-fangled order of cheap artificial teaching tricks, and entrust it to the eternal powers of nature herself."* He became the first applied educational psychologist, insisting that children begin with the concrete object before moving on to the underlying abstract concepts.

Pestalozzi emphasized the individual, encouraging spontaneity and self-activity. His students were not given preset problems with ready-made answers but were encouraged to pursue their own curiosity. He also believed in creating a nurturing environment for students, abolishing in his school the then-common practice of flogging. And he worked hard to remove the "verbosity of meaningless words" from his system, preferring to emphasize concrete observation—a doctrine he called *Anschauung* (loosely translated as "sense perception" or "object lessons"). Yet the *Anschauung* must be bolstered with concrete action, Pestalozzi cautioned: "Life shapes us and the life that shapes us is not a matter of words but action." The best way to achieve that action, he believed, was through repetition—not rote memorization, but mastering the action through practice within the context of the concrete object.

I inadvertently adopted several elements of Pestalozzi's method in my own adventures with calculus. For one thing, there was no flogging. For another, I avoided sources that relied too heavily on technical jargon—the "verbosity of meaningless words"—because I spent far too much time translating the

---

* England had its own tyrant of method and correctness: a seventeenth-century math teacher named Edward Cocker, author of a 1667 textbook called *Cocker's Arithmetick: Being a Plain and Familiar Method Suitable to the Meanest Capacity for the Full Understanding of That Incomparable Art, as It Is Now Taught by the Ablest Schoolmasters in City and Country*. This book became the standard for British grammar schools for generations. These days we have the far more succinct *Math for Dummies*.

terminology and not enough grappling with the essential concepts.

But the real key lay in the connections I was able to draw between the abstract equations and real-world examples. Don't get me wrong: Mastering the abstraction is absolutely critical to fully grasping calculus; it's just easier to see how the principles are applied if they are presented in many different familiar contexts. It's the *connection* between the abstract and concrete that eludes most students. Until I had that mimetic moment—a realization that *this* abstract equation is connected to *that* real-world example—my understanding remained incomplete, even if I managed to crank out the "correct" answer to a textbook problem.

How did I make that critical connection? By observing the world around me and then by reinforcing that observation through practice (action). I abandoned the assigned problems in standard calculus textbooks and followed my curiosity. Wherever I happened to be—a Vegas casino, Disneyland, surfing in Hawaii, or sweating on the elliptical in Boesel's Green Microgym—I asked myself, "Where is the calculus in this experience?"

The process of devising my own problems, rather than relying on existing ones, gave me insights into the discipline I would not have gained otherwise. It's akin to taking apart a mechanical toy and figuring out how to put it back together again: That process teaches you more about how that toy works than simply reading a description about its operation. I still had to do the repetitive work to hone those nascent skills and make the lesson "stick," but the repetitive process made more sense to me because it had a recognizable context.

It also helped me to see the hidden connections between

seemingly unrelated phenomena. For instance, I never realized that an exponential decay curve can describe the rate at which a cup of coffee cools, and the rate at which wet clothing dries, as well as certain processes in astronomy, economics, and population dynamics. Those very different things nonetheless are related mathematically; they are described by the same kinds of equations. If you don't "speak math," it is much more difficult to see those connections.

Two years after beginning my journey, I can't honestly say I love calculus, certainly not the way I love physics. It's more of a grudging appreciation for the role calculus plays in describing our world. I am far from mathematically fluent: As with any foreign language, that fluency comes with years of practice and regular immersion in this brave new world. I only went from the equivalent of baby talk to sounding out "See Jane run." But I have learned the history, the concepts, and the basic terminology and processes of calculus, which in turn have greatly enhanced my grasp of certain conceptual nuances in physics. More important, I am no longer reluctant to confront a simple equation, because I know it will yield a useful insight. The knee-jerk negative reaction and crippling fear are gone. And who knows? Learning is a lifelong process, so it's possible that as I continue to dabble over time, mathematics will nudge its way further into my heart.

How did I become convinced that calculus was beyond my ken? No doubt part of it stems from gender bias. There is a well-documented prejudice against women in math and science dating back thousands of years, although history gives us the rare exception, such as the plucky Sophie Germain. Such women often have been dismissed as mere statistical anomalies, but evidence is mounting that there is no innate difference in the mathematical ability of girls and boys. Any gap in performance is due primarily to sociological factors. This is a controversial statement. We would prefer to believe that the overt sexism in math and science is a thing of the past, but the reality is that these attitudes persist, even in this enlightened age.

A geometry teacher tells the entire class that the girls will probably do worse in his course because they lack spatial reasoning ability. A guidance counselor shunts female students into "practical math" classes where they learn how many ham slices each guest would need at a wedding. A physics professor insists on checking his female students' work before they can leave the lab, yet doesn't feel the need to check the work of his male students. A computer science professor dismisses any questions from female students as "lazy little-girl whining." And a calculus teacher thinks it's perfectly appropriate to measure his female students' bodies and use those measurements as part of his volume calculations in class. One woman told of her high school math teacher who made the three female students sit in the front row, "because girls have a harder time with math than boys do." It was really a flimsy excuse to ogle their cleavage and brush his crotch up against them suggestively during exams. "Guess which three people in that class were not about to be stuck in a basement computer lab with that dude?" she asked (rhetorically).

I never experienced anything so horrific; my math teachers were kind and, if not openly encouraging, they certainly were not discouraging or hostile, nor was I ever sexually harassed. My parents were supportive of my intellectual pursuits, if a bit bemused by my headier inclinations. Nobody ever told me explicitly that girls weren't as good as boys at math, yet somehow I absorbed that message anyway. Carol Tavris, a cognitive psychologist and author of several popular books (*The Mismeasure of Woman* should be mandatory reading for young women), explained to me that there are subtle, situational social cues that seep into our consciousness as if by osmosis, even if we never encounter overt negative messaging about gender.

The phenomenon is known in psychological circles as stereotype threat, and it has been confirmed in more than a hundred scientific articles. For example, a 2007 study in *Psychological Science* found that female math majors who viewed a video of a conference with more men than women reported feeling less desire to participate in the conference and less of a sense of belonging than female math majors who viewed a gender-balanced version of the video. The male math majors were immune to those subtle situational cues. That's stereotype threat in a nutshell.

These pressures are very real. Yet I can't blame my ambivalence entirely on gender. After all, plenty of boys struggle with math, too. How we self-identify in our mathematical ability sets in at an early age and colors our perception from then on. "If ever I had an Achilles heel, mathematics would surely be it," says Brian, who is studying to be an evolutionary biologist. Yet he keeps running afoul of the dreaded math classes and worries that his failures therein will dash his hopes of a career in science. "Nothing makes my blood run cold like an indecipherable word problem, and the very term 'calculus' is enough to give me nightmares," he confesses, sounding just like many of the female students I encountered.

Tavris bemoans our fascination in the United States with the notion of innate ability as the source of this kind of negative self-identification. We are born with certain built-in talents, this reasoning goes; you either have a gift for math or you don't, and no amount of hard work can make up for that lack of innate ability. I certainly bought into this notion, assuming that because it didn't come as easily to me as verbal skills, I lacked the "gift" of manipulating numbers. Yet it merely required a bit more effort on my part to learn the foreign language of mathematical symbols (vocabulary) and processes (the rules of grammar) until I became sufficiently conversant to solve basic problems. At heart, it is a foreign-language problem: Many students also struggle to learn French or German or Egyptian hieroglyphics.

Consider Deborah, whose fourth-grade teacher held multiplication table competitions in class. Deborah was highly competitive, so she worked very hard on memorizing her multiplication tables and practicing at home. As a result, she excelled in these competitions and became known as being "good at math." This had a significant impact on her later on: Whenever she struggled with an especially tough problem, she pushed through, thinking, "I should be able to do this because I'm good at math." Yet her belief in her innate ability, and her success at math, were actually the product of a lot of hard work and repeated positive reinforcement in the classroom.

Tavris also believes that American culture has an unhealthy attitude toward failure. It is considered a shameful thing rather than a natural stage of the learning process. Calla initially failed high school algebra. It shattered her confidence and instilled the telltale dislike of math that such failure so often brings. "I hated math for making me feel stupid, and because there was nothing enjoyable about it," Calla said. "It was just there, like a big black wall I would run into every once in a while, not letting me know why it was there or why I should I care." In reality, failure is how we learn. Take away the freedom to fail, and it is no wonder our students aren't learning. Science, too, relies on failed experiments and null results just as much as its justly touted successes in order to advance human knowledge.

The good news is that, regardless of the combination of factors that conspire to discourage any given individual from pursuing math and science, one good teacher can make up for all of it. I had Alan and Sean. Calla had a dedicated high school math teacher who literally changed her life. Everything changed when she took a class taught by a young woman who emphasized hands-on demonstration and applications for the math. It took some time for Calla to work through her mental blocks, but that teacher patiently guided her every step of the way with all kinds of creative approaches. They hammered away at the big black wall together until Calla finally broke through and realized she was "good" at math. She went on to major in physics in college.

There are many excellent high school math teachers, laboring in the trenches for very little pay and even less appreciation. But they are fighting an uphill battle. The way calculus is so often taught is clearly not reaching a substantial fraction of students; more often than not, like my high school self, they end up solving problems by rote, with little comprehension of why they must perform these tasks—or get so frustrated at their inability to solve problems that they reject mathematics for the rest of their lives.

Every teacher I know is heartened whenever they see that light bulb of genuine comprehension turn on in a student's brain: "Oh! *This* is *that*!" In the same way that our favorite works of art, literature, music, or theater tend to be those with elements we recognize and can respond to emotionally, we tend to respond more to books, lectures, or classroom curricula that enable us to make similar connections between the abstract concepts of math and physics and our real-world experiences. If our emotions are engaged, even better: That excitement and enthusiasm serve to fuel students' desire to persevere past the inevitable frustrating roadblocks in the quest for knowledge.

Actor David Krumholtz plays a brilliant young mathematician on the hit TV series *Numb3rs*, and he bravely participated in a panel discussion at the 2006 meeting of the American Association for the Advancement of Science on the challenge of changing negative public perceptions of math and science. With disarming frankness, he readily admitted—before a roomful of scientists—that he had flunked algebra twice in high school.

Numb3rs demonstrates the relevance of mathematics better than any pedagogical method I've yet encountered. Week after week, Charlie Epps (Krumholtz) helps his FBI agent brother crack a federal case using the tools of his trade. Math is a tough sell; couching it within the familiar crime-solving framework renders its abstract concepts not only palatable to nonscientists, but downright appealing. The show's tagline sums it up perfectly: "We all use math every day." Even Krumholtz confessed to developing a fascination for Pythagoras and the Fibonacci sequence because of their prevalence in nature and art—and were it not for his role as Charlie Epps, he might never have encountered those concepts outside of the classroom. This suggests that his struggles with math weren't due to a lack of aptitude, but to how the subject matter was presented. Like many of us, he never understood why math was important or how it could possibly be of any use in our daily lives.

There is much weeping and gnashing of teeth in academic circles about the sorry state of U.S. math-and-science education.

I don't pretend to have an easy answer to a sweeping, complex problem that confounds our best educational experts. Learning is profoundly individual, and what resonates with one student might not resonate with another. How can you systematize all those individual styles? But the power of mimesis to inspire young minds should not be ignored.

Surely it is no accident that a similar interpretation of mimesis can be applied to key breakthroughs in physics: It's that same creative impulse, finding inspiration in surprising connections. Albert Einstein credited his development of the theory of special relativity to a critical insight gleaned years before, as he sat on a train moving away from the station platform—namely that he would measure time differently from within the moving train than would someone standing on the platform ("*this* is *that*").

Watching an apple fall from a tree gave Isaac Newton his critical insight into gravity and his laws of motion: He realized the apple's position, when plotted as a function of time, formed a parabolic curve, and connected motion with geometry and algebra ("*this* is *that*"). Archimedes found the solution to the problem of Hiero's golden crown while soaking in the bathtub. My own modest breakthrough came on that fateful day in Santa Barbara, when I saw the connection between an abstract calculus equation and the motion of Saturn's rings, and realized, à la Archimedes, "Eureka! *This* is *that*!"

APPENDIX 1

Doing the Math

> The only way to learn mathematics is to do mathematics. —PAUL HALMOS

> Tell me and I'll forget. Show me and I may not remember. Involve me, and I'll understand.
>
> —NATIVE AMERICAN PROVERB

So, you've made it through *The Calculus Diaries* and feel as though you're starting to get a handle on this whole calculus thing. Maybe you're even toying with the idea of delving a bit further into the topic. This appendix is here to help you take that next step. I deliberately avoided scary equations in the main text, but sooner or later one must bite the bullet and face the actual math head-on. Nothing here is intended to "teach" calculus—this is not a substitute for the experience of an actual class, textbook, and/or a private tutor—but it will give you a taste of how the concepts discussed in the text translate into the language of math. For those who really get bitten by the calculus bug and desire even more details, I recommend *The Complete Idiot's Guide to Calculus* by W. Michael Kelley.

Here are the most common terms and symbols you'll encounter; this will help you "read" basic calculus equations:

Function. The notation for a function is $f(x)$. Whenever you see this at the start of an equation, you know you're dealing with a function of some kind: For example, $f(x) = x^2$ tells you that x^2 is a function. However, just as it's possible to convey the same meaning using different words, there can be more than one way to write an equation for a function. The function above can be written more generally as $f(x) = ax^2$, with a denoting "some constant." It is also common to write $f(x)$ simply as y. In that notation, x is the "independent variable" (it can be anything) and y is the "dependent variable" (it depends on x). This is important to remember when plotting points on a Cartesian grid (see page 268).

The above function describes a parabola. The most general notation is $f(x) = ax^2 + bx + c$, where x is the independent variable, y is the dependent variable, and a, b, and c are constants. There is also the so-called vertex form: $f(x) = a(x - h)^2 + k$. The vertex of the parabola is the point where it turns, and in this format, (h, k) delineates that point.

Even though these functions seem at first glance to be different from one another, they actually all describe the same thing: a parabola. This variation can be confusing for the beginning calculus student. I found it helpful to view the different formulations for a function as synonyms: different words that describe the same thing. The shifts in the structure are akin to shifting around clauses, subjects, and predicates of sentences in grammar—there are specific rules that kick in whenever you "reword" an equation, just as there are rules of grammar for reworking the structure of a sentence. The overall meaning conveyed remains the same. The true test of mathematical fluency is the ability to see past the symbolic clutter and find the essence of a given equation. That's why simply memorizing formulas won't suffice; you have to know what they mean.

Limit. We discussed the concept of the limit in chapter 1. Per Kelley (aka Idiot Guide Extraordinaire), "A limit is the height a function *intends* to reach [on a graph] at a given x value, whether or not it actually reaches it." For instance, the limit of $f(x) = 2x + 5$ as x approaches 3 is 11. In math-ese, that sentence would be rendered thus: $\lim_{x \to 3} f(x) = 11$. The 3 is the value of x that we are approaching, $f(x)$ represents the function of interest, and 11 is the limit. In this case, the limit is simply the value of the function, but other cases are more subtle.

Sometimes the limit does not exist, most notably when a function at a given value for x does not approach a fixed number, but instead increases or decreases infinitely. The textbook example of this is the function $f(x) = \sin\left(\frac{1}{x}\right)$ when $x = 0$. No general limit exists in that case because the function wriggles back and forth on the graph (see page 266) and never settles on a definite numeric value. Then we can just say that the limit as x approaches 0 does not exist.

Derivative. The common notation for a derivative is $\frac{d}{dx}$. Derivatives arise from ratios, or the difference between two points. The top value is the change in position, say, at two different times, while the bottom value is the difference in the time. If you want to take the derivative of $f(x) = ax^2$, you would write it out like this: $\frac{d}{dx}(ax^2) = 2ax$.

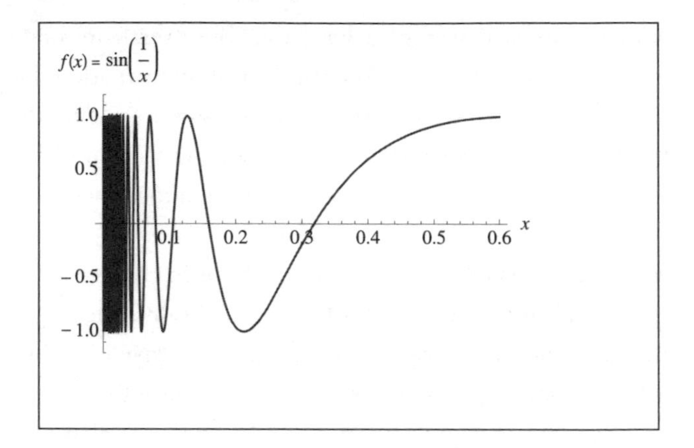

Integral. The integral is represented by a long *S*-shaped figure: $\int$·

A handy mnemonic device is to remember that integration is a process of summing (*S*), hence the elongated *S* shape is its symbol. Often, when taking an actual integral, there will be numerical values at the top and bottom of the symbol indicating the range over which one is integrating: $\int_a^b$.

This is known as a definite integral; if there is no specified range, that is called an indefinite integral. If you wanted to undo the work of the derivative on $f(x) = ax^2$, you would take an integral and write it like this: $\int ax^2\,dx = \frac{1}{3}ax^3$.

Exponentials and Logarithms. It's worth including a short note about exponentials and logarithms, which play an important role in calculus. Like the derivative and the integral, exponentials and logarithms are flip sides of the same coin: Each undoes the work of the other. Start with a number, take its exponential, and then take the logarithm of the result, and you will end up with your original number.

That original number is the *base*; to take an exponential, you multiply the base by itself x number of times. The number of times you multiply it by itself is the *power*, represented in superscript: for example, 10 multiplied by itself 5 times would be written as 10^5. When the base is 10, you can also think of the power as denoting the number of zeros to the right of the initial 1. So an exponential function would be something like 2^x, or 5^x, where the exponent is the variable. A power would be something like x^2, x^5, or x^3, where the base is the variable. It's an important distinction.

Since taking a logarithm undoes the work of the exponential, in general, the logarithm is just the number of digits in that number. Just as with exponentials, if we're dealing with a perfect power of 10, for example, the logarithm is the number of zeros to the right of the initial 1: $\log(10) = 1$, $\log(100) = 2$, $\log(1{,}000) = 3$, and so on. Or, to put it as generally as possible, $\log(10^x) = x$. The only catch is that you can't take the logarithm of a negative number: no such animal exists. The logarithm inverts the exponential, but you can't *get* a negative number with exponentials.

THE PLOT THICKENS

Back at the start of my foray into calculus, my physicist spouse, Sean, would leave simple problems on our home whiteboard for me to solve, like little mathy love notes. (Yes, we have a whiteboard at home. Doesn't everyone?) The first set of problems focused on learning how to plot the points generated by specific functions onto a Cartesian grid, then connecting the dots to see the shape of the resulting curve (or "face" of the function). I quickly figured out this was much easier to do in a handy program called Grapher: You just plug in different values for the variable(s) in a given function, hit Return, and the correct curve miraculously appears. (You can do the same thing in Excel.)

It's fun to play with Grapher, but frankly, I found it just as instructive to slowly plot out a few functions by hand. Many of us have difficulty grasping the notion of just what a function is: The textbook definitions, while technically correct, usually convey little actual meaning to nonmathy sorts like me. Literally taking a given function apart, point by point, and slowly rebuilding it again can help bridge that gap in communication.

Let's plot the function $f(x) = ax^2$ onto a Cartesian grid with the familiar x and y axes. Remember that $f(x)$ is just another way of writing y for calculus purposes; so we're working with $y = ax^2$. The process is simple, if tedious. Assuming that $a = 1$, all we are doing is plugging in different values for x to get the corresponding value for y and plotting the point where they intersect onto the grid. I found it helpful to write down those initial values into columns first.

If x =	**Then y =**
–5	25
–4	16
–3	9
–2	4
–1	1

0	0
1	1
2	4
3	9
4	16
5	25
. . .	. . .

We already know this will be a parabola. I chose whole numbers, both positive and negative, for simplicity's sake, but you can plug in any value for x along the real number line: positive, negative, fractions, and so on. (If you don't include negative values, you only get half the parabolic curve.) Remember that the function technically comprises *all* possible values for x in that equation taken together—i.e., an infinite number of values. That would be tedious to plot indeed. But you can plug in enough values along the number line, plot out the corresponding points on the grid, and at some point you accumulate enough points that a definite curvy pattern emerges when you connect the dots.

I've described curves as representing the "faces" of functions, but those faces can have multiple expressions. Someone who is happy, sad, or angry will have the same basic features, but their faces can look quite different depending on the emotions they are experiencing. The same is true of functions. For instance, the constant a in our equation determines the size and direction of the parabola. The larger the value of a, the steeper, or thinner, the resulting parabola will be. Also, if a is positive, the parabola opens upward; if a is negative, it opens downward. Where $a = 2$, we get a parabola that looks like the one on page 270.

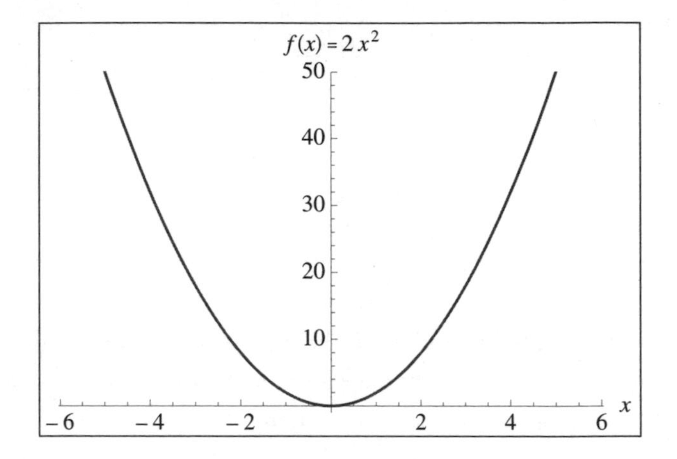

Where $a = -2$, we get the exact same parabolic curve, only inverted (falling below the x axis) because the sign is now negative:

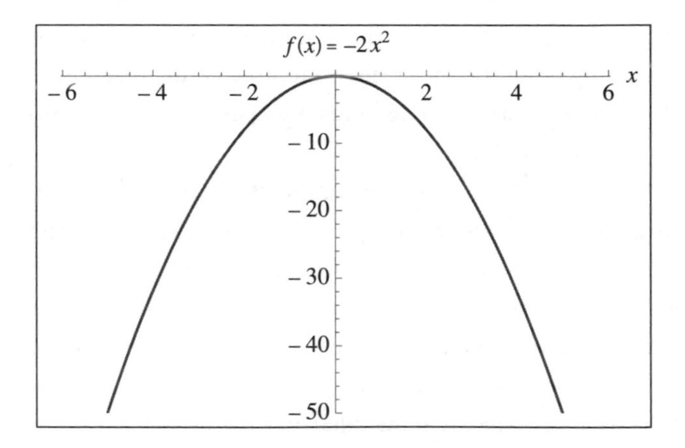

Finally, we can add additional variables: $f(x) = ax^2 + bx + c$, also known as $y = x^2 + bx + c$. It's fun to play with the basic equation and see firsthand how changing each value for the different variables is reflected in the shape of the resulting curve. For instance, this is what you get when you plug in the values $a = 3$, $b = 8$, and $c = 10$:

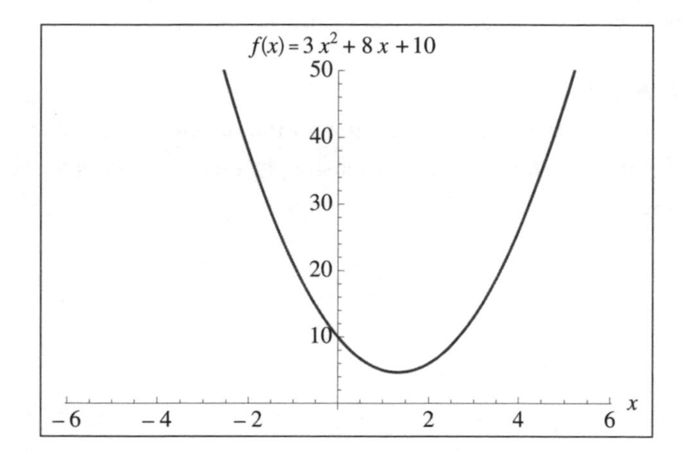

It's still the same basic function; the fundamental nature of its "face" hasn't changed, it's just expressing different "emotions."

TOP TEN FUNCTIONS

While it's useful to practice graphing a few functions by hand, certain functions crop up so frequently that it's worth committing their "faces" (curves) to memory. The top ten most common functions are listed below. They should already be somewhat familiar, since you've encountered all but one (the logarithm) in the text.

For good measure, I'm also including their derivatives and integrals, because it's important information for any beginning calculus student, and why do the work of crunching those numbers all over again when past generations of mathematicians have done it for you? It will also help you to see the connection between the two in practice, namely, how the derivative undoes the work of the integral, and vice versa.

1. A Constant: $f(x) = c$

This is the function you'd use for the velocity of a car moving at a constant speed down a straight road, for example, as discussed in chapter 2.

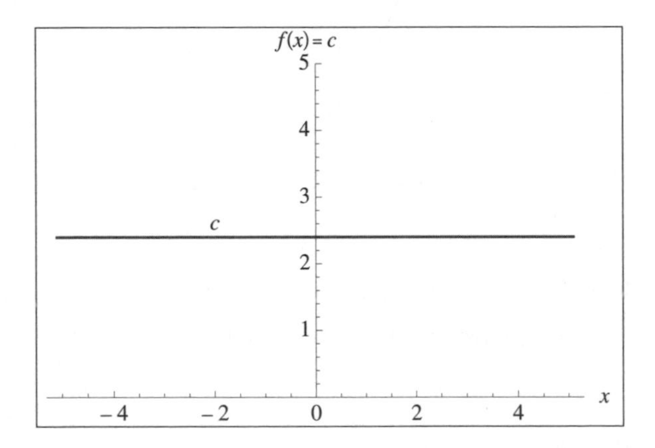

Derivative:

$$\frac{d}{dx}c = 0$$

The notation to the left of the equal sign tells us we are taking a derivative of the constant c. The answer is 0 because the derivative measures a rate of change. A constant, by definition, does not change, so the rate (and hence the derivative) is 0.

Integral:

$$\int c dx = cx$$

Here, the notation to the left of the equal sign tells us we are taking an integral. Remember that the integral is the flip side of the derivative. If we take a derivative of the velocity to determine the acceleration of a car moving at a constant rate, then we take an integral of the velocity to determine how far we traveled between our starting point (a) and ending point (b). The c tells us that we are dealing with a constant, and the dx tells us we are taking an integral of the derivative of that constant.

If we were taking a definite integral, we would write this differently: $\int_a^b c\,dx = (b-a)c$. The a and b variables at the top and bottom of the integral sign simply define the range over which we are taking the integral. On the right side of the equal sign, the notation simply tells us that we are subtracting our starting position (a) from our ending position (b) and multiplying by the constant c to determine how far we traveled.

2. A Straight Line: $f(x) = ax + b$

This is the function you'd use for the velocity of a car accelerating at a constant rate, for example, also discussed in chapter 2.

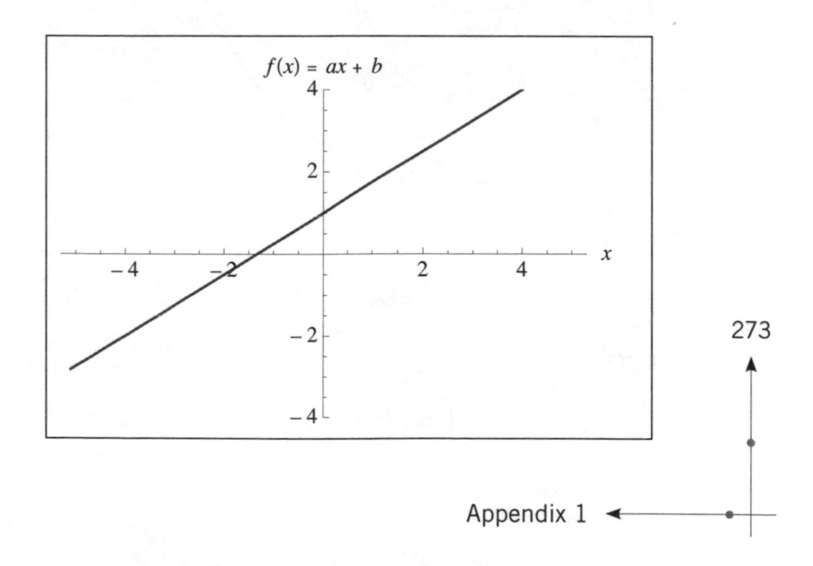

Derivative:

$$\frac{d}{dx}(ax+b)=a$$

Integral:

$$\int (ax+b)\,dx=\frac{1}{2}ax^2+bx+c$$

3. A Parabola: $f(x) = ax^2$

This function pops up all over the place in physics, whether we're dealing with the trajectory of a cannonball, the acceleration of a falling apple, or our motion (changing position with respect to time) on the Tower of Terror free-fall ride in chapter 4.

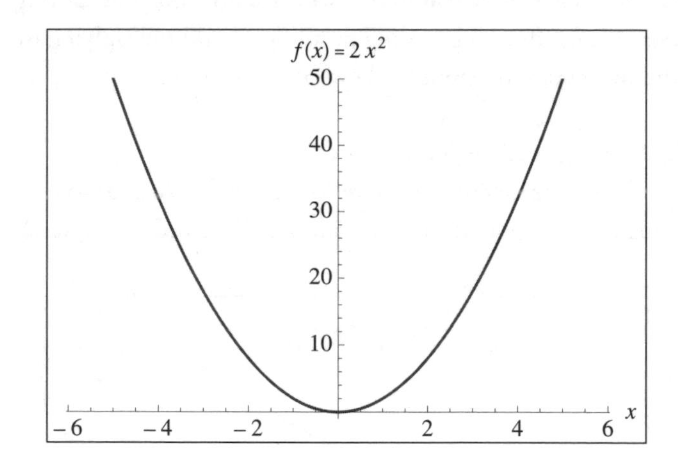

Derivative:

$$\frac{d}{dx}(ax^2)=2ax$$

Integral:

$$\int ax^2\,dx=\frac{1}{3}ax^3+c$$

4. Exponential Growth Curve: $f(x) = 10^{ax}$

We covered the basics of exponentials earlier. For an exponential function, we fix the base number and let the power to which it is raised be the variable: In this case, the base is 10 and the power is *ax*. This is the function we would use to describe the almost certain annihilation of the human race by voracious zombies in chapter 6 or the rapid growth rate of the Dutch tulip trade in chapter 5.

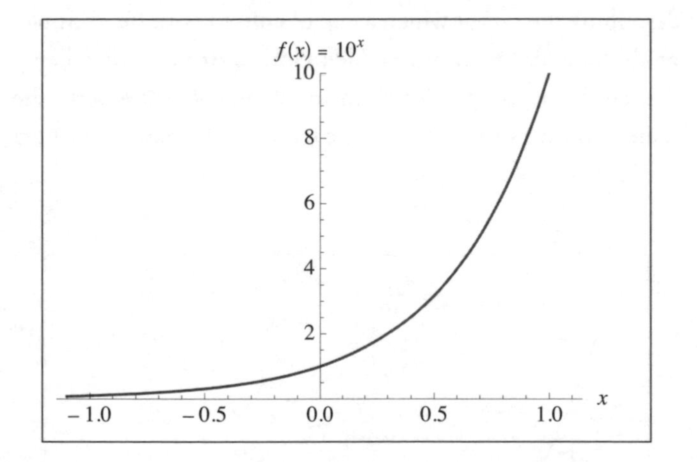

Derivative:

$$\frac{d}{dx}10^{ax} = (a \log\ e)10^{ax}$$

Integral:

$$\int 10^{ax}\,dx = \frac{1}{a \log\ e}10^{ax}$$

You'll notice that there is some new notation here: log *e*. This means the logarithm of Euler's constant (*e*). I didn't discuss Euler's constant specifically in the text, despite its importance, because, frankly, it muddies the waters of comprehension for those dipping a toe into calculus for the first time. It is an irrational number, like π, which means it goes on forever when written out in explicit form: $e = 2.71828\ldots$ That's why it is usually just left as e in an equation. The logarithm of e, in case you're wondering, is 0.43429 . . .

5. Exponential Decay Curve: $f(x) = 10^{-ax}$

This is another function that pops up frequently in physics, describing the rate at which a cup of coffee cools, for example, or the rate at which our sodden clothes dry out after being drenched on Splash Mountain in chapter 4. It's exactly the same as the exponential growth curve, but the power to which the base is raised is negative.

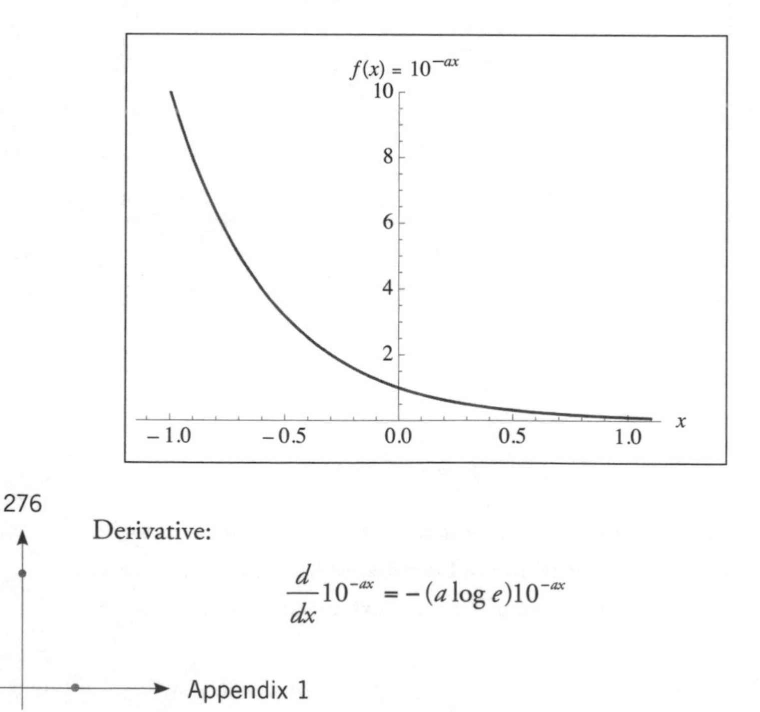

Derivative:

$$\frac{d}{dx}10^{-ax} = -(a \log e)10^{-ax}$$

Integral:

$$\int 10^{-ax}\,dx = \frac{-1}{a \log e}10^{-ax}$$

Note that the derivative and integral of the exponential decay curve also are virtually identical to that of the exponential growth curve, except for the minus sign in the power.

6. Logarithm: $f(x) = \log(ax)$

We didn't discuss the logarithmic function specifically in the main text, but this is what physicists often use to determine the entropy (disorder) of a physical system, such as a box filled with gas, a black hole, or Carnot's heat engine in chapter 7. Note that because there is no such thing as a logarithm for a negative number, the curve is not defined for negative values of x. Instead, as x approaches 0 moving from the right, the logarithm goes to minus infinity.

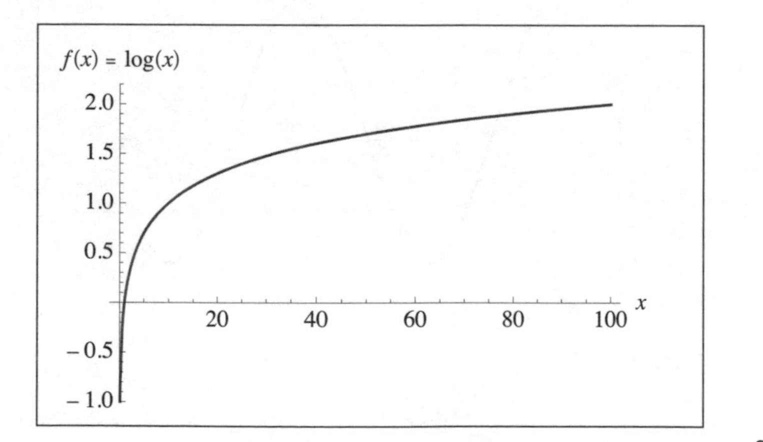

Derivative:

$$\frac{d}{dx}\log(ax) = \frac{\log e}{x}$$

Integral:

$$\int \log ax\, dx = x\log(ax) - x + c$$

7. Sine: $f(x) = \sin(ax)$

This is an example of a periodic function: one whose values repeat over and over, at the same rate and at the same intervals in time. That interval is called the period. We encountered sine waves, or sinusoid curves, in chapter 9 while talking about ocean waves, but the concept can apply to any wavelike phenomenon (light waves, sound waves, gravitational waves) or any process that repeats itself after a fixed period of time (the ticking of a clock, a human heartbeat, the rising of the sun every twenty-four hours).

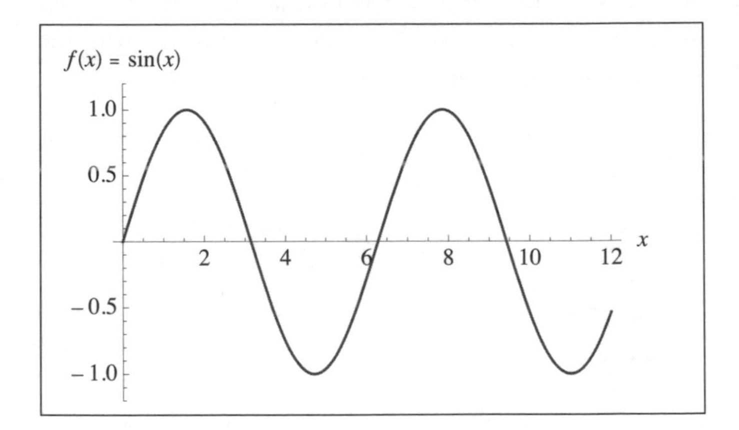

Derivative:

$$\frac{d}{dx}\sin(ax) = a\cos(ax)$$

Integral:

$$\int \sin ax\, dx = \frac{-1}{a}\cos(ax)$$

8. Cosine: $f(x) = \cos(ax)$

The cosine is the complement to the sine function, and is also an example of a sinusoid curve, applying to wavelike behavior.

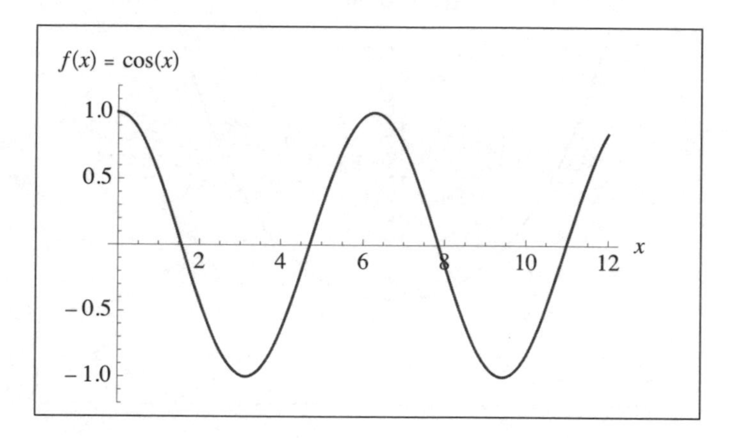

Derivative:

$$\frac{d}{dx}\cos(ax) = -a\sin(ax)$$

Integral:

$$\int \cos ax\, dx = \frac{1}{a}\sin(ax)$$

9. Catenary (or Hyperbolic Cosine):

$$\mathrm{f}(x) = \cosh x = \frac{e^{x} + e^{-x}}{2}$$

This is the curve we discussed in chapter 8 that when inverted describes the strongest possible shape for an arch. Here we encounter Euler's constant again, this time as the function e^x. Like other irrational numbers, e has some unusual properties. For instance, the function e^x is the only function—other than $f(x) = 0$—that is equal both to its own derivative and to its own integral. You can see this clearly in the notation below.

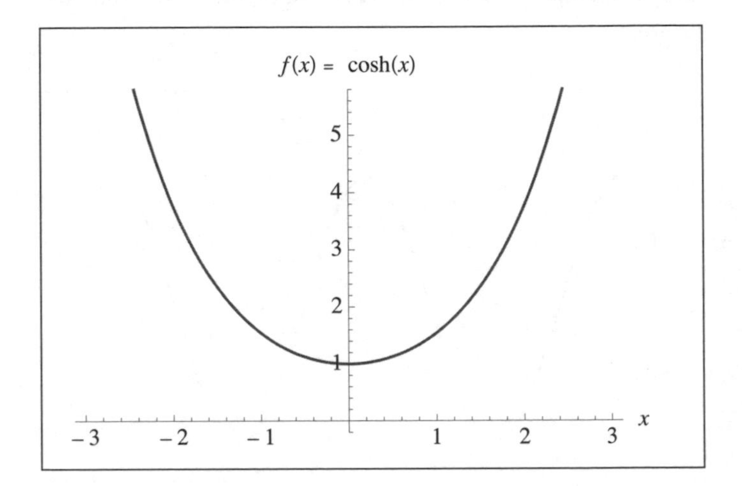

Derivative:

$$\frac{d}{dx}(\cosh x) = \frac{1}{2}(e^{x} - e^{-x})$$

Integral:

$$\int \cosh\, dx = \frac{1}{2}(e^{x} - e^{-x})$$

10. Bell Curve (Gaussian Distribution): $\mathrm{f}(x) = ae^{-x^2}$.

This is perhaps the function best known to the general populace, albeit one that is often misunderstood. We encountered it in chapter 3 when discussing the probabilities of craps, but it is applicable to almost any situation involving a large number of random variables, such as the Black-Scholes model used in economics for options pricing, among other applications. It is also useful for calculating the probability of a given characteristic in a large population and for determining SAT scores or academic grades (known as "grading on a curve").

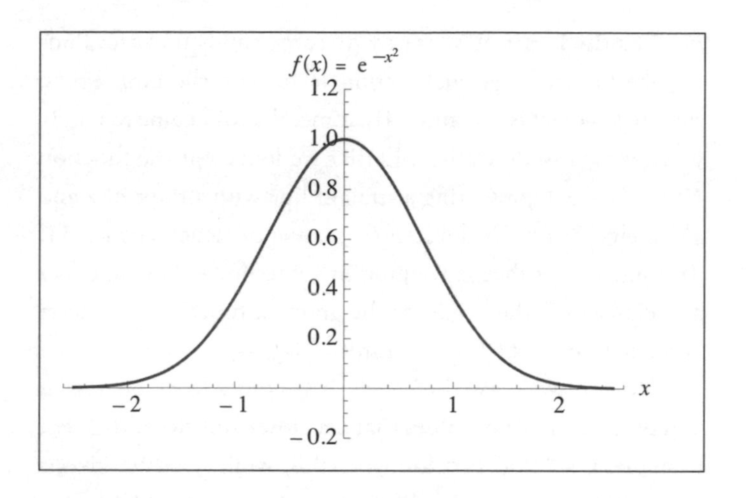

Derivative:

$$\frac{d}{dx}e^{-x^2} = -2xe^{-x^2}$$

Integral: There is no known integral for the Bell curve. It can be calculated on a computer but not written in an explicit form.

WORKING IT OUT

Now it's time to put all the pieces together and see how calculus really works. These are simple examples that can be done with pencil and paper, but it's worth investing in a scientific calculator if you're planning to delve deeper into calculus. Let the machines do the tedious task of number crunching; *real* math is all about solving problems creatively, not rote mechanics.

Finding the Limit. We'll start with some handy tricks for finding the limit of a given function (assuming the limit exists; sometimes there is no limit). Trust me, this will come in handy when we get to derivatives. Earlier we looked at the function $f(x) = 2x + 5$, representing a straight line with a slope of 2 and a y-intercept of 5. The limit of $f(x)$ as x approaches 3 equals 11. This just means that as we plug in values for x that are closer and closer to 3, the height of the graphed function gets closer and closer to $y = 11$ (aka the limit).

How do we know this? Well, it becomes fairly obvious if you plug in a series of values that get closer and closer to 3. For example, $x = 2.9$ gives a limit of 10.8, while $x = 2.95$ gives a limit of 10.9, and $x = 2.99999$ gives a limit of 10.99998. The closer the value of x is to 3, the closer the answer is to 11. Ergo, 11 is the limit of this particular function when $x = 3$.

But this is a tedious and time-consuming process that merely approximates the limit; we'd prefer to determine the limit precisely. The simplest strategy is called the substitution method: You just plug in the value of whatever number is specified under the "lim" notation. For example, let's find the limit of a parabolic function, $f(x) = x^2$, as x approaches 2: $\lim_{x \to 2} x^2$. Plug 2 into the equation, and we get 4. So $\lim_{x \to 2} x^2 = 4$.

Similarly, to find the answer to $\lim_{x \to 4} (x^2 - x + 2)$, make $x = 4$, so that $4^2 - 4 + 2 = 14$. So $\lim_{x \to 4} (x^2 - x + 2) = 14$.

You can verify this by using the graph of the function $f(x) = (x^2 - x + 2)$: another parabola. Simply plug in a few values both above and below 4, and you should see the results come closer and closer to 4 as those values trend closer and closer to 4.

Alas, it is not always that simple. Sometimes when you substitute the number specified under the "lim" notation, you get a nonsensical result, such as a 0 in the denominator, which is a mathematical taboo. In that case, you could use the factoring method to simplify things a little. Let's say we want the answer to $\lim_{x \to -3} \frac{x^2 - 9}{x + 3}$. If we try to plug the value –3 into the equation, we end up with $\frac{0}{0}$. This is not helpful.

So we switch tactics and factor the numerator; x^2 and 9 both happen to be perfect squares. (There's a reason I chose this particular example.) The result is $\lim_{x \to -3} \frac{(x+3)(x-3)}{x+3}$.

Aha! We learned in high school algebra that if you have the same expression in the numerator and denominator, they cancel each other out: In this case, we have $(x + 3)$ in both the numerator and denominator. Cross them out, and that gives us a far simpler problem: $\lim_{x \to -3} (x - 3)$.

Now we can revert back to the substitution method and plug in –3. This time we get: –3 – 3 = –6. So $\lim_{x \to -3} \frac{x^2 - 9}{x + 3} = -6$. As Sean explains, "The limit of the function is well defined at $x = -3$, even though the function itself is not."

Finding the Slope of a Straight Line. In chapter 2 we went on a road trip from Los Angeles to Las Vegas, using highly idealized parameters to illustrate the fundamental concepts of what is essentially precalculus. For illustrative purposes, we'll use another idealization here: that of a car accelerating to the speed limit, then traveling for a while at a constant speed, before braking suddenly to avoid an obstacle in the road. If we graphed our changing velocity as a function of time, the resulting curve would look like this:

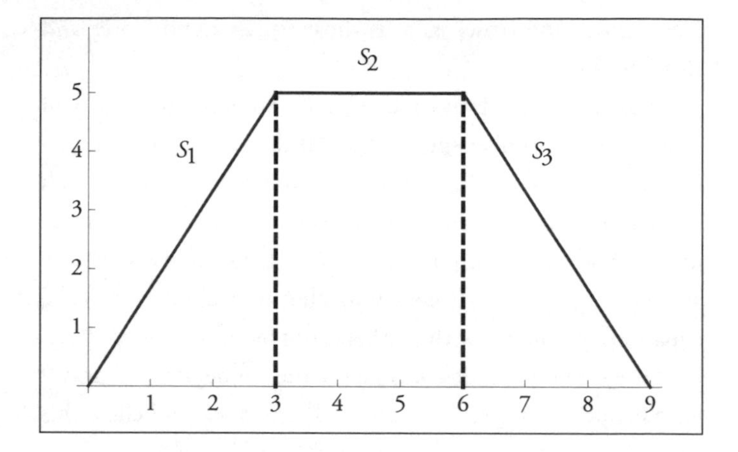

Yes, this shape technically is still a "curve," despite its straight edges. Because we are dealing with straight lines, it is pretty easy to determine the slopes of S_1, S_2, and S_3. We simply pick any two points at random on the line of interest, (a,b) and (c,d), and plug those values into this handy formula: $s = \frac{d-b}{c-a}$. For S_1, let's pick the beginning and ending points (0,0) and (3,5). Plug those values into our formula and we get $s = \frac{5-0}{3-0} = \frac{5}{3}$. It's simple arithmetic to determine that $S_1 = 1$. We can follow the same process for S_3, using points (6,5) and (9,0). We get $s = \frac{0-5}{9-6} = \frac{-5}{3}$. The fact that the slope is negative means we were slowing down.

The slope of S_2 is 0 because it is perfectly horizontal, or flat. Because we are traveling at a constant speed, there will be no difference between the values of those two points. (Perfectly *vertical* lines have no slope at all, and thus the slope is said to be "undefined.")

Recall that the steepness of the slope tells us the rate at

which those values are changing (the derivative); the steeper the slope, whether trending upward or downward, the faster its value is changing. You can also see this trend in the above formula. If the numerator (top) is larger than the denominator (bottom), then the y's are changing faster and therefore the line is getting steeper. If the denominator is larger, that means the x's are changing faster and the line forms a shallow incline, because it is moving more quickly to left or right than it is moving up and down. So it should be clear that the above curve describes a car accelerating, then cruising at a constant speed before decelerating.

Finding the Area. It is an equally simple matter to find the area under this particular curve by breaking it into common geometric shapes: a rectangle bounded by two triangles.

We can find the area of the two triangles by halving the base and multiplying that number by the height, written algebraically as $A = \frac{1}{2} bh$. (A stands for area in this context, b stands for base, and h stands for height.) We find the area of the rectangle by multiplying the width times the height, written algebraically as $A = wh$, where w stands for width. Then it's just a matter of adding those three areas together to find the total area under our simple curve.

For both triangles, $h = 5$ and $b = 3$. So we can multiply 2.5 by 3 to get an area of 7.5 for each. The same goes for the rectangle in the center; we multiply 3 by 5 to get 15. Then we add it all together (15 + 7.5 + 7.5) and we end up with a total area of 30. Simple, right?

THINGS GET MESSY

Alas, the real world rarely fits neatly into these sorts of idealized models. In reality, for the above example, our speed and direction would be varying constantly, and we would not be dealing with simple straight lines, but with curves. This is the true value of calculus: It helps us solve more difficult problems dealing with change and motion using known derivatives and integrals for given functions. Once again, the derivative describes rates of change and corresponds to the slope of the tangent line to a particular point on the curve, while the integral corresponds to the area under a curve. It's just a little trickier to find those values when dealing with irregular geometric shapes.

In chapter 4, we experienced free fall while riding the Tower of Terror and learned that plotting our motion (change in position) as a function of time onto a graph produced a parabolic curve. This parabolic curve represents our position function (height, h, as a function of time, t): $h(t) = \frac{1}{2}at^2 + b$. Let's say we want to figure out our instantaneous velocity at a specific point. We need to find the slope of the tangent line for that point, which is equivalent to the derivative of our position function.

So we know our position function, and we also know the value of the constant b, namely our starting height. The Tower of Terror is 199 feet high; we can round up to an even 200 feet to make our calculations easier. So $b = 200$. Finally, we know the value of a, since falling objects travel at -32 feet per second per second; half of that is -16. (The sign is negative because our height decreases as we fall.) Plug those values into our starting function and we get $h(t) = -16t^2 + 200$. Now we're ready to start differentiating.

Derivatives: The Hard Way. I won't lie to you: Things are about to get ugly. But it's instructive to walk through every painful step just so we can fully appreciate how useful calculus can be when we see the simplified process in the *next* section. We begin by picking our point (h_1, t_1). We draw a straight line tangent to that point, and now we want to find the slope, which in turn will give us our instantaneous velocity. The problem is that our chosen tangent line only hits the curve on a single point. We can still pick another nearby point (h_2, t_2) and use our nifty formula above to calculate the difference between them, but this time it won't be the exact slope, merely an approximation. The closer the two points are to each other, the better the approximation we will get. (Note that when both points (h_1, t_1) and (h_2, t_2) are the same, we end up with $\frac{0}{0}$. Having 0 as your denominator is taboo in math.)

But we want to find the *exact* value for the slope of the tangent line. The good news is that we have another nifty formula for just this sort of problem: $\frac{\Delta h}{\Delta t} = \frac{h_2 - h_1}{t_2 - t_1}$, where Δ stands for a tiny increment. The bad news is that we have no idea what the values are for h_2, h_1, or Δh.

First things first: We need to find the value of h_2. We can find h_2 by plugging t_2 into our starting function, like this: $h(t_2) = -16t_2^2 + 200$.

We know that $t_2 = t_1 + \Delta t$, so we can substitute that expression for t_2, like this: $h(t_2) = -16(t_1 + \Delta t)^2 + 200$.

We want to break this down as much as possible. Think of it as deconstructing our equation, i.e., reducing it to its individual components, the better to manipulate the pieces. For instance, we can rewrite the equation above as

$$h(t_2) = -16(t_1 + \Delta t)(t_1 + \Delta t) + 200.$$

I'll skip over the next few steps, which just involve further deconstruction according to the "grammatical rules" of math; suffice it to say, we end up with this:

$$h(t_2) = -16(t_1{}^2 + 2t_1\Delta t + (\Delta t)^2) + 200.$$

Things are starting to get very confusing, and we're not done yet. Next we need to find the value of h_1. Happily, this is just our starting function: $h_1 = -16t_1{}^2 + 200$.

Now we can subtract h_1 from h_2 to get Δh:

$$\Delta h = -16t_1^2 - 32t_1\Delta t - 16(\Delta t)^2 + 200 - (-16t_1^2 + 200)$$

Once we're done deconstructing that equation and canceling out all the extraneous stuff, we end up with a far more malleable version: $\Delta h = -32t_1\Delta t - 16(\Delta t)^2$.

Finally, we divide the whole mess by Δt: $\frac{\Delta h}{\Delta t} = \frac{-32t_1\Delta t - 16(\Delta t)^2}{\Delta t}$. This can be simplified even further to give us $\frac{\Delta h}{\Delta t} = -32t_1 - 16\Delta t$.

Now our old friend the limit comes into play. We take the limit by setting Δt to 0: $v(t_1) = \lim_{\Delta t \to 0} \frac{\Delta h}{\Delta t}$.

We've already solved for $\frac{\Delta h}{\Delta t}$ above, so we can plug that value in and rewrite this as $\lim_{\Delta t \to 0}(-32t_1 - 16\Delta t) = -32t$.

When all is said and done, we end up with $v(t) = -32t$. Physics fans will recognize this as the stock formula for determining velocity: Velocity is equivalent to acceleration multiplied by time, written generically as $v = at$.

Derivatives: The Easy Way. I hope you feel a few twinges of compassion for the poor souls who went through the above process over and over again to find the derivatives of all the major functions, then compiled them into a master list for subsequent generations. It's so much easier these days, because we *know* the derivative of t^2 is $2t$, for example, and even if we don't, we can look it up.

Let's revisit our problem again using this simpler process. We know we have a starting function of $h(t) = -16t^2 + 200$. We want to know our instantaneous velocity when $t = 1$ second. So we take a derivative to find the relevant velocity function: $v(t) = \frac{dh}{dt}$.

Because we know the value of h, we can use substitution to rephrase the question as $\frac{dh}{dt} = \frac{d}{dt}(-16t^2 + 200)$.

Next we break up those parenthetical expressions to get

$$\frac{dh}{dt} = \frac{d}{dt}(-16t^2) + \frac{d}{dt}(200).$$

Now we move the –16 from the parentheses to get $\frac{dh}{dt} = -16\frac{d}{dt}(t^2) + 0$. Where did that 0 come from? We took the derivative of 200, which is a constant, and one of the hard-and-fast rules in calculus is that the derivative of any constant is 0, because the derivative describes the rate of change and a constant doesn't change.

All we have left to do is find the derivative of t^2. This is where we can skip all those in-between steps, because we know that the derivative of t^2 is $2t$. That means we can plug $2t$ into the equation, so we end up with $\frac{dh}{dt} = -16(2t)$.

We know that $\frac{dh}{dt}$ is equivalent to $v(t)$. Multiply it out, and you end up with the same answer we got via our earlier belabored process: $v(t) = -32t$. Translated back into plain English: Our instantaneous velocity at our chosen point on the curve is –32 feet per second and because the velocity is equivalent to the slope of our tangent line, that slope is also $-32t$. (Our velocity is changing over time, so t must be included.)

Taking an Integral. The integral is the reverse of the derivative, so now we will reverse the question. This time we know our velocity as a function of time: $v = at$. We want to determine our position at a given point in time, denoted as $h(t)$. The integral corresponds to the area under a curve, which is fairly easy to calculate in this case, because our velocity function translates graphically into a straight line. So we just need to find the area under that line by dividing the base by 2 and multiplying that number by the triangle's height $\left(\frac{1}{2}bh\right)$.

But let's say our starting function gives us a bona fide curve with no straight lines or triangles to assist us; now things become complicated. We've already seen a method for approximating the area under a curve in chapter 1: the aptly named method of exhaustion pioneered by Eudoxus, whereby we fill in the curve with a series of rectangles for which it is a simple matter to determine the area. We calculate those individual areas, then add them all together to get an approximation of the area under the curve. The smaller the rectangles we use, the more of them it takes to fill the area under the curve, and the closer the approximation. We literally could do this forever, using infinitesimally small rectangles.

Luckily for us, there is another way: taking an integral. It's a bit harder than finding the area of a triangle, but it simplifies matters greatly when trying to determine the area under a curve, so it's worth walking through the process for $v = at$. (I will spare you the full derivation. You're welcome.)

We can write out our question mathematically like this: $h(t) = \int v(t)\,dt$. This just says that we are integrating velocity (v) over time (t), adding all those instantaneous velocities together to determine our position (h).

Thanks to our handy velocity function, we know that $v = at$, so we can replace $v(t)$ with at to get $h(t) = \int at\,dt$.

Another handy rule of calculus is that whenever you integrate a constant multiplied by a function, like *at*, you can bring that constant outside the integral symbol, like this:

$$h(t) = a\int t\,dt.$$

Now we can get rid of both our integral symbol and the *dt* by looking up the integral for *t*, which turns out to be $\frac{1}{2}t^2 + c$. We've picked up a constant because of another hard-and-fast rule of calculus: Whenever you take an indefinite integral—i.e., when no beginning and ending point is specified—your answer is going to have a constant (hence the waggish habit of physicists to jokingly add "plus a constant" to random observations). It makes sense if you think about it for a moment: The integral corresponds to the area under a curve, which by definition describes a given range. Even if we don't know what that range is, we still need a placeholder in our equation: *c* represents that constant of unknown value.

We plug all of that into our equation to get this:

$$h(t) = a(\frac{1}{2}t^2) + ac.$$

Look familiar? It's our best buddy, the parabola! So now we know that if our velocity increases like a straight line ($v = at$), our position increases like a parabola (also written as $h(t) = \frac{1}{2}at^2 + c$). And we can prove it mathematically.

Fun with Functions. But the point is, we've found our position function in just a few easy steps: $h(t) = \frac{1}{2}at^2 + c$. Now we can determine our position (*h*) for any value of *t*.

Remember that in our free-fall scenario, $b = 200$ and $a = -32$. For instance, what is our position (h) when $t = 1$? It's 184 feet. When $t = 2$, $h = 136$ feet, when $t = 3$, $h = 56$ feet, and so on. In fact, we can devise an algebraic equation from our position function to determine when h will equal 0 and we will hit the ground if we fell from atop the Tower of Terror. Since we're solving for t, it looks like this: $t = \sqrt{-\frac{2b}{a}}$.

Plug those numbers into our formula like this: $t = \sqrt{-\frac{2(200)}{-32}}$. The minus signs cancel out and we get $t = \sqrt{\frac{400}{32}}$.

Now it's just a matter of factoring down until we get $t = \frac{5}{\sqrt{2}}$. Our handy calculator tells us the square root of 2, we divide 5 by that, and the answer is $t = 3.5$. So we will hit the ground and go splat 3.5 seconds after we start falling.

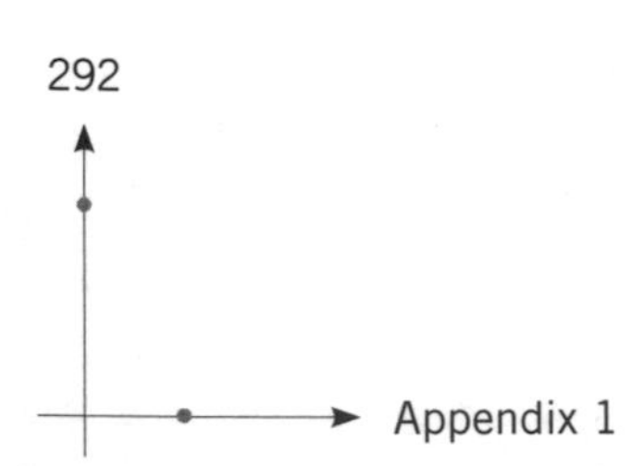

APPENDIX 2

Calculus of the Living Dead

Time to nut up or shut up.

—TALLAHASSEE,
Zombieland

A particularly virulent form of human-adapted mad cow disease sweeps across the United States in the 2009 hit film, *Zombieland*, transforming most of the nation's populace into ravenous zombies. The film follows a ragtag group of unlikely survivors on a road trip in hopes of finding someplace yet untouched by the disease, ending with a pitched battle against zombie hordes in an abandoned amusement park. *Zombieland* beat out the *Dawn of the Dead* remake as the top-grossing zombie film to date. You just know there's going to be a sequel.

Let's say that in this much-anticipated sequel, Columbus, Wichita, Little Rock, and the Twinkie-craving Tallahassee manage to find an uninfected haven and enjoy a brief respite from battling the Undead. Then the first zombies appear, and the refugees know their days of peace are numbered. If they knew the rate of infection—that is, how quickly the zombie population is growing—they could predict when those numbers would become overwhelming and could plan to evacuate before the situation grew dire. With luck, they could just keep moving, always staying one step ahead of the zombie plague. All they need is a bit of calculus.

To solve this problem, we must delve into the murky realm of differential equations, which sound scary but are really just equations that contain a derivative. This is a bit more convoluted than the problems in appendix 1, but as I discovered in my own odyssey, at some point you've got to nut up or shut up and face the monsters head on if you're serious about learning calculus. You never know when this sort of thing might help you survive a zombie apocalypse.

Why do we need a differential equation? Remember that exponential growth (or decay) in populations—how fast something grows (or declines)—depends on the size of the population itself. (A perfect exponential model would require infinite resources, a condition that rarely exists, but for illustrative purposes, it will suffice.) Solving the differential equation will give us the key to determining how many zombies there will be at any time. We'll use a textbook sample problem, cribbed from Kelley: $\frac{dy}{dx} = ky$.

In the above equation, *y* represents the population of zombies, *x* represents the time that has passed, and the derivative is the rate of change in the number of zombies. The constant (*k*) describes how quickly the zombies multiply.

The first step toward solving a differential equation usually involves shifting different variables to opposite sides of the equation—we're essentially rephrasing the question. In this case, we want to isolate *y* on the left side of the equation. A mathematician will tell you that $\frac{dy}{dx}$ isn't "really" a fraction, but standard rules of algebra still apply. To move *y* from left to right, we must divide both sides of the equation by *y*: do this and you get $\frac{dy}{y(dx)} = k$.

We want *y* all by itself, so now we have to move *dx* to the right side of the equation. This is easily accomplished by multiplying both sides of the equation by *dx* to get $\frac{dy}{y} = kdx$.

Since we want to add up the number of zombies over time, we now integrate both sides of the equation to find the integral. We would write this mathematically as $\int \frac{dy}{y} = \int kdx$. Fortunately, *k* is a constant, and we learned a neat trick for determining the integral of a constant in chapter 2. Just as the integral of 5 is 5*x*, the integral of *k* is *kx*. Meanwhile, on the left-hand side, we have the integral of $\frac{1}{y}$ with respect to y, which just happens to be the natural logarithm function $\ln(y)$. We would end up with $\ln(y) = kx + c$.

Because we used an indefinite integral, we've picked up a constant (c), as well as that pesky natural log function (ln). Fortunately, we can cancel out the natural log function by using the natural exponential function, e^x, described in appendix 1. This is the inverse function to a natural log, which means it undoes what the log has done. (Inverse functions are tools to eliminate something you don't want in an equation—in this case, the natural log function.) We can rewrite our equation like this: $e^{\ln(y)} = e^{kx+c}$. All we've done is change each side of the prior equation so that it's expressed as a power of e. Because e^x and $\ln x$ cancel each other out, we end up with $y = e^{kx+c}$.

This is a solution of sorts, but we can simplify it even further. First, there's a basic rule when dealing with exponentials that when we have x^{a+b}, we can rewrite it as $x^a \cdot x^b$. So we can rephrase the equation yet again as $y = e^{kx} \cdot e^c$.

Finally, we can rewrite e^C as simply C since both are constants; that lets us combine them into one big constant. After all is said and done, we end up with the function $y = Ce^{kx}$. Whew! This is the equation that encodes the answer to the question we've posed.

If you're anything like me, by now your head hurts, and you just want to be done with it. But remember the zombies! We need to figure out how fast the zombie infection is spreading, because our very lives may depend on it. As we learned in chapter 2, the solutions to differential equations aren't specific numbers; they are new equations. And this particular equation holds the key to determining the growth rate of the zombie population.

What does our new equation tell us? Well, C stands for the initial population of zombies (a constant number that doesn't change), while y now stands for the total zombie population after a certain amount of time (denoted by x) has passed. We've got Euler's constant (e) lurking around as well, but we need not worry about it just yet. That leaves k, which will tell us the rate of zombie infection.

We need to figure out the value of k. We start with the initial zombie population: Let's say on day 1 there were 19 people who ate contaminated hamburger and turned into zombies ravenous for tasty brains. Ten days later, we count again and discover their ranks have swelled to 193 zombies. That's all the data we need to solve for k, using our handy little formula: $y = 193$ (new number of zombies), $C = 19$ (beginning number of zombies), and $x = 10$ (days that have passed): $193 = 19e^{10k}$.

Next we take a series of steps to simplify and solve this equation. Since k is the value we want to find, we must isolate k on the right side of the equation. First we divide by 19 on both sides of the equation: $\frac{193}{19} = e^{10k}$.

Now we need to get rid of that annoying exponential function, which we do by reintroducing the natural logarithm: $\ln\left(\frac{193}{19}\right) = 10k$.

Next we divide both sides of the equation by 10 to isolate k: $\frac{\ln\left(\frac{193}{19}\right)}{10} = k$.

Finally, we get our answer: $k = 0.2318$. Aren't you glad you invested in that calculator? Brace yourself, because we're not quite done yet.

Now we can plug the values for C and for k into our handy little equation: $y = 19e^{0.2318x}$. With this information, we can determine the number of zombies there will be *any* number of days in the future, just by varying the *x* factor. Voila! We have a truly predictive model; that is the beauty of the mathematical function.

For instance, how many zombies will there be after thirty days? Make $x = 30$, and we get 19,914 zombies. Hmmm. That's some serious exponential growth. I'm sure our merry band of zombie killers would agree: We will be outnumbered very quickly. Evacuation is definitely in order.

BIBLIOGRAPHY

BOOKS AND ARTICLES

Abraham, Marc. "Think of a Number: A Multitude of Math Sins." *Guardian*, April 28, 2009.

Ashford, Karen. "Cedar Point Physics Outing Lauded." Canada.com, April 26, 2008.

Bardi, Jason Socrates. *The Calculus Wars: Newton, Leibniz, and the Greatest Mathematical Clash of All Time*. New York: Thunder's Mouth Press, 2006.

Bell, E. T. *Men of Mathematics*. New York: Simon & Schuster, 1937.

Beller, Peter C. "Fill 'Er Up with Human Fat." *Forbes*, December 21, 2008.

Berlinski, David. *A Tour of the Calculus*. New York: Pantheon, 1995.

Berzon, Alexandra. "The Gambler Who Blew $127 Million." *Wall Street Journal*, December 5, 2009.

Bland, Eric. "Web-Crawling Program ID's Disease Outbreaks." *Discovery News*, July 18, 2008.

Bohannon, John. "Social Science: Tracking People's Electronic Footprints." *Science* 314, no. 5801 (November 10, 2006): 914.

———. "Friends or Acquaintances? Ask Your Cell Phone." ScienceNOW, August 17, 2009.

Bressoud, David M. *The Queen of the Sciences: A History of Mathematics* (DVD). Chantilly, VA: Teaching Company, 2008.

Chang, Kenneth. "Study Suggests Math Teachers Scrap Balls and Slices." *New York Times*, April 25, 2008.

Cheever, Henry T. *Life in the Sandwich Islands: The Heart of the Pacific as It Was and Is*. Ann Arbor: University of Michigan Library, 2005.

Clawson, Calvin. *Conquering Math Phobia: A Painless Primer*. New York: Wiley, 1991.

Cowen, Tyler. *Discover Your Inner Economist*. New York: Dutton, 2007.

Danzig, Tobias. *Number: The Language of Science*. New York: Penguin/Plume, 2007.

Darbyshire, Charles. *My Life in the Argentine Republic 1852–1894*. London: F. Warne & Co., 1917.

Dash, Mike. *Tulipomania: The Story of the World's Most Coveted Flower and the Extraordinary Passions It Aroused*. New York: Crown, 2000.

Devlin, Keith. *The Language of Mathematics: Making the Invisible Visible*. New York: Henry Holt, 2000.

———. *The Unfinished Game*. New York: Basic Books, 2008.

Devlin, Keith, and Gary Lorden. *The Numbers Behind Numb3rs*. New York: Plume Books, 2007.

Ellenberg, Jordan. "We're Down $700 Billion, Let's Go Double or Nothing: How the Financial Markets Fell for a 400-Year-Old Sucker Bet." Slate.com, October 2, 2008.

Finkelmeyer, Todd. "Culture, Not Biology, Key Factor to Math Gender Gap, UW Researchers Say." Madison.com, June 1, 2009.

Gleason, Alan, trans. *Who Is Fourier? A Mathematical Adventure*. Boston: Language Research Foundation, 1995.

Goodwin, Liz. "Monsters vs Jane Austen." DailyBeast.com, September 5, 2009.

Gosline, Anna. "The Calculus of Saying I Love You." Inkling.com, October 13, 2007.

Grahame-Smith, Seth. *Pride and Prejudice and Zombies*. Philadelphia: Quirk Books, 2009.

Harford, Tim. *The Undercover Economist: Exposing Why the Rich Are Rich, the Poor Are Poor—and Why You Can Never Buy a Decent Used Car*. New York: Oxford University Press, 2006.

Hempel, Sandra. *The Strange Case of the Broad Street Pump: John Snow and the Mystery of Cholera*. Berkeley/Los Angeles: University of California Press, 2007.

Hernandez, Nelson. "The Thrills of Physics: For High School Students, Theme Park Becomes a Laboratory." *Washington Post*, April 26, 2008.

Houston, James D., and Ben Finney. *Surfing: A History of the Ancient Hawaiian Sport*. New York: Pomegranate Communications, 1996.

Hsu, Jeremy. "Second Life Bank Crash Foretold Financial Crisis." MSNBC.com, November 21, 2008.

Hurt, Jeanette. "Patience Is a Virtue." *Hemispheres*, May 2009.

Johnson, Steven. *The Ghost Map: The Story of London's Most Terrifying Epidemic—and How It Changed Science, Cities and the Modern World*. New York: Riverhead, 2006.

Kaminski, J. A., V. M. Sloutsky, and A. F. Heckler. "The Advantage of Abstract Examples in Learning Math." *Science* 320, no. 5875 (2008): 454–55.

Kelley, W. Michael. *The Complete Idiot's Guide to Calculus*, 2d ed. New York: Alpha Books, 2006.

Kemp, Martin. "Inverted Logic: Antoni Gaudi's Structural Skeletons for Catalan Churches." *Nature* 407, no. 838 (October 19, 2000).

Kestenbaum, David. "A Prayer Book's Secret: Archimedes Lies Beneath." NPR.org, July 27, 2006.

Kojima, Hiroyuki. *The Manga Guide to Calculus*. San Francisco: No Starch Press, 2009.

Kolata, Gina. *Flu: The Story of the Great Influenza Pandemic of 1918 and the Search for the Virus That Caused It*. New York: Touchstone, 2001.

———. "Putting Very Little Weight in Calorie Counting Methods." *New York Times*, December 20, 2007.

———. *Rethinking Thin: The New Science of Weight Loss—and the Myths and Realities of Dieting*. New York: Farrar, Straus & Giroux, 2007.

Krendl, Anne C., Jennifer A. Richeson, William M. Kelley, and Todd F. Heatherton. "The Negative Consequences of Threat: A Functional Magnetic Resonance Imaging Investigation of the Neural Mechanisms Underlying Women's Underperformance in Math." *Psychological Science* 19, no. 2, 168–75.

Laak, Phil. "Kelly's Criterion." *Bluff*, April 2008.

Lehrer, Tom. "The Derivative Song." *American Mathematical Monthly* 81 (May 1974): 490.

Levenson, Thomas. *Newton and the Counterfeiter: The Unknown Detective Career of the World's Greatest Scientist*. New York: Houghton Mifflin Harcourt, 2008.

Lofgren, E. T., and N. H. Fefferman. "The Untapped Potential of Virtual Game Worlds to Shed Light on Real World Epidemics." *The Lancet Infectious Diseases* 7 (2007): 625–29.

London, Jack. *The Cruise of the Snark* (1911). New York: Narrative Press, 2001.

Lovell, Michael C. *Economics with Calculus*. Hackensack, NJ: World Scientific Publishing, 2004.

Lucibella, Mike. "The Best Approach for Avoiding Zombies," Inside Science News Service, September 28, 2009.

Mackay, Charles. *Extraordinary Popular Delusions and the Madness of Crowds* (1841). New York: Three Rivers Press, 1980.

Mankiw, N. Gregory. *Principles of Microeconomics*, 3d ed. Mason, OH: Thomas-Southwestern, 2004.

Maor, Eli. *e: The Story of a Number*. Princeton, NJ: Princeton University Press, 1994.

McKellar, Danica. *Math Doesn't Suck*. New York: Penguin Books, 2007.

Mlodinow, Leonard. *The Drunkard's Walk: How Randomness Rules Our Lives*. New York: Pantheon, 2008.

Munz, Philip, et al. "When Zombies Attack! Mathematical Modeling of an Outbreak of Zombie Infection," in *Infectious Disease Modeling Research Progress*. Haupauge, NY: Nova Science, 2009, pp. 133–50.

Mythbusters. "Ancient Death Ray," airdate September 29, 2004. Silver Spring, MD: Discovery Communications.

———. "Archimedes' Death Ray," airdate January 25, 2006. Silver Spring, MD: Discovery Communications.

Nadeau, Robert. "The Economist Has No Clothes." *Scientific American*, March 25, 2008.

Nitta, Hideo, and Keita Takatsu. *The Manga Guide to Physics*. San Francisco: No Starch Press, 2009.

O'Rourke, P. J. *On the Wealth of Nations*. New York: Atlantic Monthly Press, 2007.

Ouellette, Jennifer. "Architects Bridge the Void." *New Scientist*, June 13, 2006.

Paulos, John Allen. *Innumeracy*. New York: Farrar, Straus & Giroux, 1988.

———. *A Mathematician Reads the Newspaper*. New York: Random House, 1995.

Peeples, Lynn. "Eco-Conscious Gyms Allow Members to Spin Calories into Electricity." Scienceline.org, February 24, 2009.

Pestalozzi, J. *How Gertrude Teaches Her Children: An Attempt to Help Mothers to Teach Their Children and an Account of the Method* (London, 1805), trans. L. Holland and F. Turner. Syracuse, NY: Swann Sonnenschein, 1894.

Pool, Bob. "Getting the Slant on L.A.'s Steepest Street." *Los Angeles Times*, August 21, 2003.

Poundstone, William. *Fortune's Formula: The Untold Story of the Scientific Betting System That Beat the Casinos and Wall Street*. New York: Hill & Wang, 2005.

Reymeyer, Julie. "A Prayer for Archimedes." *Science News*, October 8, 2007.

Rodriguez, Linda. "A Brief History of Dubious Dieting." *Mental Floss*, January 28, 2009.

Rosenwald, Michael S. "For Hybrid Drivers, Every Trip Is a Race for Fuel Efficiency." *Washington Post*, May 26, 2008.

Sample, Ian. "Doubt Cast on Archimedes' Killer Mirrors." *Guardian*, October 24, 2005.

Schepisi, Fred, dir. *I.Q.* Paramount Pictures, 1994.

Seife, Charles. *Zero: The Biography of a Dangerous Idea*. New York: Penguin, 2000.

Severson, Kim. "Seduced by Snacks? No, Not You." *New York Times*, October 11, 2006.

Smith, Reginald. "The Spread of the Credit Crisis: View from a

Stock Correlation Network." http://arxiv.org/abs/0901.1392, January 12, 2009.

Smith?, Robert, et al. "The OptAIDS Project: Towards Global Halting of HIV/AIDS." *BMC Public Health*, November 18, 2009.

Soltis, Greg. "Worms Do Calculus to Find Food." LiveScience .com, July 23, 2008.

Squires, Sally. "Bringing Nutrition Home." *Washington Post*, February 12, 2008.

Starbird, Michael. *Calculus Made Clear* (DVD). Chantilly, VA: Teaching Company, 2001.

Stein, Jeanine. "Do the Math, Lose the Weight." *Los Angeles Times*, May 11, 2009.

Stokstad, Erik. "Americans' Eating Habits More Wasteful Than Ever." *Science*, November 25, 2009.

Thomas, Nicholas. *Cook: The Extraordinary Voyages of Captain James Cook*. New York: Walker, 2004.

Thomson, Helen. "What's Luck Got to Do with It? The Math of Gambling." *New Scientist*, August 11, 2009.

Tobias, Sheila. *They're Not Dumb, They're Different: Stalking the Second Tier*. Tucson: Research Corporation, 1994.

———. *Overcoming Math Anxiety*. New York: Norton, 1995.

Twain, Mark. *Roughing It* (1872). New York: Signet Classics, 2008.

Van Hensbergen, Gijs. *Gaudi: A Biography*. New York: HarperCollins, 2001.

Vastag, Brian. "Virtual Worlds, Real Science: Epidemiologists, Social Scientists Flock to Online World." *Science News*, October 27, 2007.

Venkatraman, Vijaysree. "An Electric Workout Through Pedal Power." *Christian Science Monitor*, November 13, 2008.

Wansick, Brian. *Mindless Eating: Why We Eat More Than We Think*. New York: Bantam/Dell, 2006.

Ward, Mark. "Deadly Plague Hits Warcraft World." BBC News, September 22, 2005.

———. "Virtual Game Is a 'Disease Model.'" BBC News, August 21, 2007.

Yan, Zhenya. "Financial Rogue Waves." http://arxiv1.library.cornell.edu/abs/0911.4259?context=q-fin.CP, November 22, 2009.

Zaslavsky, Claudia. *Fear of Math: How to Get Over It and Get on with Your Life*. New Brunswick, NJ: Rutgers University Press, 1994.

Zwillinger, David. *Standard Mathematical Tables and Formulae*, 31st ed. New York: Chapman & Hall, 2002.

GENERAL ONLINE RESOURCES

Sites:

Albion Research
American Dietetic Association
Anecdotage.com
Arcytech.com
ArXiv
BBC News Online
The Calculus Page
Cracked.com
CrapsCenter.com
eHow.com
Health Map
How Stuff Works
Investopedia
Los Angeles Wheelmen
MacTutor History of Mathematics Archives, University of St. Andrews (Scotland)
Mouse Planet
National Curve Bank
Reducingstereotypethreat.org
The Straight Dope
Surfing Legends

Surfing for Life
U.S. Centers for Disease Control
Walkability Score
Wikipedia
William Poundstone.net
Wired.com
Wolfram MathWorld
World Health Organization

Blogs:

3 Quarks Daily
Aetiology
BldgBlog
Built on Facts
Calculated Risk
Cocktail Party Physics
Confounded by Confounding
Cosmic Variance
Damn Interesting
Doctor Housing Bubble
Effect Measure
The Frontal Cortex
Good Math, Bad Math
Gravity and Levity
In the Dark
Io9
Laelaps
Mike the Mad Biologist
Neurophilosophy
Not Exactly Rocket Science
Physics arXiv Blog
The Quantum Pontiff
Retrospectacle
Sciencegeekgirl
Southern Fried Science
Starts with a Bang
Swans on Tea
Tiny Cat Pants
Uncertain Principles

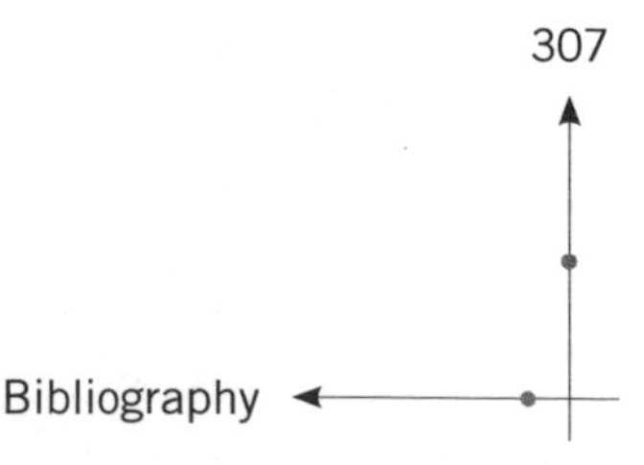

INDEX

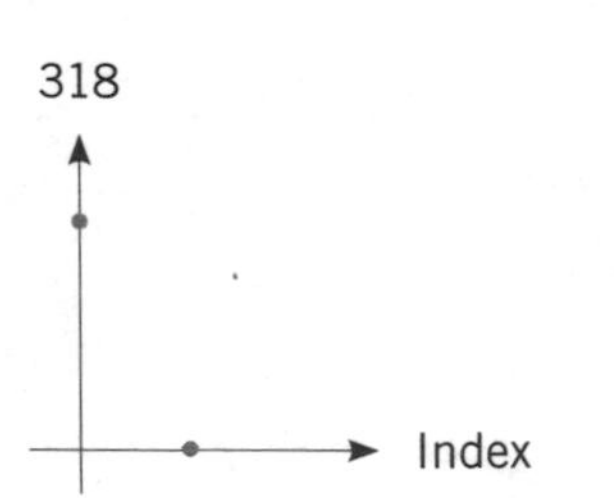
Index